Advances in Image and Data Processing using VLSI Design, Volume 1

Smart vision systems

IOP Series in Next Generation Computing

Series editors

Prateek Agrawal
University of Klagenfurt, Austria and Lovely Professional University, India

Anand Sharma
Mody University of Science and Technology, India

Vishu Madaan
Lovely Professional University, India

About the series

The motive of this series is to develop a trusted library on advanced computational methods, technologies and their applications.

This series focuses on the latest developments in next generation computing, and in particular on the synergy between computer science and other disciplines. Books in the series will explore new developments in various disciplines that are relevant for computational perspective including foundations, systems, innovative applications and other research contributions related to the overall design of computational tools, models and algorithms that are relevant for the respective domain. It encompasses research and development in artificial intelligence, machine learning, block chain technology, quantum cryptography, quantum computing, nanoscience, bioscience-based sensors, IoT applications, nature inspired algorithms, computer vision, bioinformatics etc. and their applications in the areas of science, engineering, business and social sciences. It covers a broad spectrum of applications in the community, from industry, government, and academia.

The aim of the series is to provide an opportunity for prospective researchers and experts to publish works based on next generation computing and its diverse applications. It also provides a data-sharing platform that will bring together international researchers, professionals and academics. This series brings together thought leaders, researchers, industry practitioners, and potential users of different disciplines to develop new trends and opportunities, exchange ideas and practices related to advanced computational methods and promote interdisciplinary knowledge.

Advances in Image and Data Processing using VLSI Design, Volume 1

Smart vision systems

Edited by
Sandeep Saini
LNM Institute of Information Technology, Jaipur, India

Kusum Lata
LNM Institute of Information Technology, Jaipur, India

Abhishek Sharma
LNM Institute of Information Technology, Jaipur, India

G R Sinha
Myanmar Institute of Information Technology, Mandalay, Myanmar

IOP Publishing, Bristol, UK

ISBN 978-0-7503-3919-3 (ebook)
ISBN 978-0-7503-3917-9 (print)
ISBN 978-0-7503-3920-9 (myPrint)
ISBN 978-0-7503-3918-6 (mobi)

DOI 10.1088/978-0-7503-3919-3

Version: 20211201

IOP ebooks

British Library Cataloguing-in-Publication Data: A catalogue record for this book is available from the British Library.

Published by IOP Publishing, wholly owned by The Institute of Physics, London

IOP Publishing, Temple Circus, Temple Way, Bristol, BS1 6HG, UK

US Office: IOP Publishing, Inc., 190 North Independence Mall West, Suite 601, Philadelphia, PA 19106, USA

Dedicated to my wife and daughter.
Sandeep Saini

Dr Sharma would like to thank his parents and family for their patience, encouragement, and kind support. Thanks to Manu, Madhu, and Mihika for everything.
Abhishek Sharma

Dedicated to my late grandparents, my teachers and the Revered Swami Vivekananda.
G R Sinha

Contents

Preface

Image-processing-based applications and systems are becoming an integral part of modern smart cities, systems, and daily routine tasks. The technical details of the design and implementation of such systems are very important to understand signal and image-processing tasks and their hardware implementations. This book discusses the design frame, modeling concepts, and applications of image-processing-based systems using very large scale integration (VLSI) design techniques. Considering the vast range of applications, we propose to publish this book in two volumes. Each volume will focus on specific topics and contain the chapters related to those topics. In volume 1, we focus on object detection, recognition, smart traffic management, surveillance systems, face detection, gesture-based automated systems, smart cities based on automated cameras, and further similar applications that are based on the concepts of image and video processing and finally implemented in hardware for optimum performance. In volume 2, we focus on biomedical applications, in which image-processing-based hardware design is used for applications such as cancer detection, ECG and EEG measurements, medical-imaging-based healthcare systems, and 'smart' homes for elderly and disabled people.

Volume 1 covers 'smart' computer vision and imaging systems that further highlight various aspects of smart vision, such as object detection, object tracking, pattern recognition, smart traffic monitoring systems, surveillance systems, face detection, gesture-based automated systems, smart city-based systems using automated cameras and smart sensors, and many more similar applications based on the concepts of image and video processing, implemented in hardware for optimum performance. The chapter-wise description of volume 1 of the book is as follows: in chapter 1, Sandeep *et al* present an field-programmable gate array (FPGA) implementation of a 1024 bit Rivest–Shamir–Adleman (RSA) cryptosystem on the Virtex-5 XC5VLX2110T-2FF1738 FPGA board. Finite State Machine (FSM) architectures for key generation, encryption, and decryption are proposed. Encryption requires the computation of the modular exponentiation of 1024 bits with accuracy and efficiency, and the repeated modular multiplication technique carries it out. For decryption, an L-R binary approach is used, which deploys a modular multiplication module. Our design achieves efficiency terms of the throughput/area ratio, whose value is 6.36 megabits per second (Mbps)/slice for encryption and 1.66 Mbps/slice for decryption, which is better than existing implementations described in the literature with the same key size. Furthermore, encryption and decryption of a gray image are demonstrated with the help of the Xilinx System Generator (XSG) model. Furthermore, to ensure efficient encryption of the shared key by RSA, histograms of the key images before and after encryption are generated and analyzed for encryption strength. In chapter 2, Abhishek *et al* study modern machine learning (ML) methods for object detection. We present multiple management strategies for the proper operation of convolutional neural networks (CNNs) for image classification, and we discuss various types of data timing scheme, including quantification, used to reduce computational cost and decrease memory complexity in dedicated hardware. In chapter 3, Kyle and Ali present embedded intelligence for tracking facial expressions with the aim of examining one component of facial recognition software: facial expression detection and evaluation, working to maintain a certain level of accuracy while running on hardware that is significantly weaker in computational power than a modern desktop workstation. In chapter 4, Payal *et al* discuss the role of delay and power optimization techniques in VLSI design. The chapter covers the development of new design methodologies that simultaneously reduce power, area and delay. Power dissipation does not scale linearly with delay. In chapter 5, Abhishek Bharti *et al* present the Smart Interactive Virtual Assistance System (SIVAS) (a voice user interface (VUI) application) and describe its implementation. It can effectively and efficiently translate human voice commands into computer actions.

Chapter 6 presents a topical survey of computing solutions for plant disease classification using deep learning techniques. In this chapter, Aniket *et al* survey contemporary articles related to plant recognition and the identification of diseases, based on computing solutions that use recent popular techniques, such as machine learning and deep learning. This chapter also summarizes the techniques used by existing methods, along with their major challenges, such as research gaps in this particular area. In chapter 7, Anirban *et al* present hardware intellectual property (IP) cores for image-processing functions. This chapter discusses high-level synthesis (HLS) design methodology for dedicated optimized hardware IP cores for five image-processing functions—the Gaussian blur filter, the Sobel edge detector, the Prewitt gradient edge detector, image zooming and image contrast/brightness adjustment. In chapter 8, Alpana *et al* study and suggest an efficient underwater image cryptosystem that uses a novel hybrid algorithm. The proposed algorithm is tested on various underwater images collected from the internet and compared with the original block-based scheme. After simulation experiments, it is found that the algorithm provides better contrast enhancement and reconstruction of the underwater images. In chapter 9, a survey of thresholding in image processing is presented by Eshaan. This chapter discusses various image thresholding methods, such as histogram-based methods, the entropy-based method, the clustering-based method, the local method, and the attribute-based method. This chapter also documents the mathematical formulation of the techniques used, when required. A review of the quality assessment of fruits and vegetables using near infrared (NIR) spectroscopy is presented by Mariappan *et al* in chapter 10. This chapter elaborates on the non-destructive method and the the principles of NIR spectroscopy techniques. Different spectral modes used for the assessment of fruit, i.e. reflectance, interactance, and transmittance, are explained. The measurement of different attributes of multiple kinds of fruit and vegetable are addressed using optimal variable selection and calibration methods. The main objective of this review is to described the role and importance of NIR spectroscopy in the assessment of various fruit quality attributes.

In chapter 11, Vardhana *et al* present a design and implementation of processors for secured image-processing applications. The chapter proposes the design and implementation of a high-speed 8-bit Vedic processor, in particular, the arithmetic logic unit, which is developed and built using ancient Vedic mathematical techniques. The proposed processor also includes signal-processing operations, such as convolution, fast Fourier transform (FFT), and cryptographic operations. The improvement in the efficiency in terms of speed and area is shown by comparing the results with the results of existing algorithms, in chapter 12. Premananda *et al* present a low-power modified phase-locked loop (PLL) using the adaptive voltage level at supply (AVLS) technique for biomedical applications. The objective of this work is to modify the existing PLL architecture and incorporate power reduction techniques to minimize the power consumption of the PLL. The subcomponents of the PLL, phase detector, charge pump, loop filter, voltage-controlled oscillator, and frequency divider are individually designed and optimized for power consumption and then integrated to form a PLL. Two different phase detector circuits are considered: a linear phase detector and another phase frequency detector. For the loop filter design, metal–oxide–semiconductor field-effect transistors (MOSFETs) are used instead of resistors and capacitors. In chapter 13, Lavanya *et al* present image multiplication with a power-efficient approximate multiplier using a 4:2 compressor. Approximate computing has been introduced to reduce the increased power consumption. An 8×8 unsigned Dadda multiplier is created by placing the proposed 4:2 compressor using a probabilistic pruning type of approximation technique. This Dadda multiplier is employed for the pixel-by-pixel multiplication of two images. Using results simulated by Cadence Virtuoso, it is observed that the power consumption is decreased by 82.83% when compared with the exact multiplier. Chapter 14 presents the relative learning of different procedures for video shot boundary detection (SBD) with hardware means. In this chapter, Shrikant *et al* discuss different methods which are implemented to achieve SBD along with their comparative analysis. Many advanced methodologies that use machine learning approaches and other traditional methods are framed here. The results, which were validated on TREC Video Retrieval Evaluation (TRECVID) datasets, are also compared for different approaches. In addition, challenges such as motion, illumination, and lighting effects are also discussed. Also, the hardware implementation for the same is included in the chapter, and various aspects of the FPGA-based hardware system are discussed. Finally, in chapter 15, Surbhi *et al* present hardware–software co-simulation of vehicle license plate detection on a ZedBoard system-on-chip (SoC) platform. This chapter aims to demonstrate the SoC implementation of a vehicle license plate detection system (VLPDS), which identifies license plates using a histogram-based edge processing technique. The system is designed with MATLAB Simulink and XSG and then executed on a Xilinx Zynq-7000 ZedBoard SoC. The extensive application of XSG in image processing decreases the complexity of the structural design while also allowing hardware–software co-simulation, which may be a unique feature.

Overall, this book provides information about basic and advanced image-processing techniques, how these applications are revolutionizing the modern world, how various tasks are being automated using these techniques, and why dedicated hardware is the best way to implement such applications. This book also emphasizes the real challenges in this field due to the complex and non-stationary nature of signals processed using a variety of applications for analysis, classification, and identification. The chapters can be read independently by research scholars, R&D engineers, and graduate students who wish to explore research in the field of image processing and VLSI design.

Acknowledgments

Dr Saini expresses his heartfelt appreciation to his parents Mr L C Saini and Sudesh Saini, wife Manpreet, and daughter Dhriti, for their wonderful support and encouragement throughout the completion of this important book on **Advances in Image and Data Processing using VLSI Design**.

Dr Abhishek Sharma would like to thank Manorama Sharma (wife), Madhuparna Sharma and Mihika Sharma (daughters) for their unconditional support and inspiration.

Dr Kusum would like to give special thanks to her husband Dr Nitin and the whole family for their unconditional love, support, and understanding without which completing this would not have been possible.

Dr Sinha also expresses his gratitude and sincere thanks to his family members, his wife Shubhra, daughter Samprati, parents and teachers.

We would like to thank all our friends, well-wishers, and all those who keep us motivated to do more and more, better and better. We sincerely thank all contributors for writing about the relevant theoretical backgrounds and real-time applications of the analysis of medical modalities for improved diagnoses in modern healthcare.

We express our humble thanks to John Navas and all the editorial staff of IoP Publications for their great support, necessary help, appreciation, and quick responses. Finally we want to thank, everyone, in one way or another, who helped us editing this book.

Last, but not least, we would also like to thank God for showering us with his blessings and the strength to do this type of novel and quality work.

Sandeep Saini
Kusum Lata
Abhishek Sharma
G R Sinha

Editors' biographies

Sandeep Saini

Sandeep Saini received his BTech degree in Electronics and Communication Engineering from the International Institute of Information Technology, Hyderabad, India in 2008. He completed his M.S. from the same institute in 2010. He obtained his PhD from the Malaviya National Institute of Technology, Jaipur, India.

He has been working at the LNM Institute of Information Technology, Jaipur, as an assistant professor from 2011. He has worked as adjunct faculty at the International Institute of Information Technology (IIIT), Bangalore (Deputation at Myanmar Institute of Information Technology, Mandalay, Myanmar) for two years and as a lecturer at Jaypee University of Engineering and Technology, Guna for three semesters. His research interests are in the areas of deep learning, machine learning, natural language processing, the cognitive modeling of language learning models, and biomedical and agricultural applications of deep learning. He has published one book, four book chapters, and over 30 international journals and conference papers. Sandeep has been a member of the Institute of Electrical and Electronics Engineers (IEEE) from 2009 and is also an active member of the Association for Computing Machinery (ACM).

Kusum Lata

Kusum Lata received her MTech and PhD degrees from the Indian Institute of Technology (IIT), Roorkee, India and the Indian Institute of Science (IISc), Bangalore, India in 2003 and 2010, respectively. She also worked as a research associate at the Centre of Electronics Design and Technology, IISc Bangalore, India for six months after completing her PhD. From June 2010, she worked as a lecturer for three years at the Indian Institute of Information Technology, Allahabad (IIIT-A), India. She worked as assistant professor from December 2013 to February 2016, and since March 2016 has been working as associate professor in the Department of Electronics and Communication Engineering at the LNM Institute of Information Technology, Jaipur.

She was the recipient of the *Outstanding Research Paper* award at the first *Asia Symposium on Quality Electronic Design* (ASQED-2009), 15–16 July 2009, Kuala Lumpur, Malaysia. Her research interests include digital circuit design using FPGAs, design for testability, the formal verification of analog- and mixed-signal designs, and hardware security. Kusum has been a member of the IEEE since 2003 and an active member of the ACM since 2011. She is also a lifetime member of the Computer Society of India.

Abhishek Sharma

Abhishek Sharma received his BE in Electronics from Jiwaji University, Gwalior, India and a PhD in Engineering from the University of Genoa, Italy. He is presently working as an assistant professor in the Department of Electronics and Communication Engineering at the LNM Institute of Information and Technology, Jaipur, RJ, India. He is also a member of the IEEE, the IEEE's Computer Society, and its Consumer Electronics Society; he is a lifetime member of the Indian Society for Technical Education, the Advanced Computing Society, India, and the Computer Society of India. He is also the center lead of the LNM Center of Smart Technology (L-CST). His research interests are real-time systems, the IoT, and smart embedded systems.

G R Sinha

G R Sinha (PhD, FIE, FIETE, SMIEEE) is adjunct professor at the International Institute of Information Technology, Bangalore (IIITB) and currently deputed as professor at the Myanmar Institute of Information Technology (MIIT), Mandalay, Myanmar. He was visiting professor (honorary) at the Sri Lanka Technological Campus, Colombo during 2019–2020. He was a visiting professor teaching the short graduate course on Cognitive Science and Brain Computing Research at the University of Sannio, Italy from September 2020 to March 2021.

He has published 278 research papers, book chapters, edited books and books at the international level in the areas of image processing, medical imaging, biomedical signal processing, cognitive science, optimization techniques, sensors, data science, data deduplication and outcome-based education (OBE) with globally reputed publishers, such as the IOP, Elsevier, Springer, CRC, IGI Global, and the IEEE. He is an associate editor of five reputed SCI/Scopus indexed journals and guest editor of a few journals. He has 23 years' teaching and research experience. He has been the Dean of Faculty and an executive council member of CSVTU and is currently a member of the Senate of MIIT. Since 2017, Dr Sinha has been delivering ACM lectures across the world as an ACM Distinguished Speaker in the field of DSP. His more important assignments include being the Expert Member for the Vocational Training Program of the Tata Institute of Social Sciences (TISS) for two years (2017–2019); the Chhattisgarh Representative of the IEEE MP Subsection's Executive Council (2014–2017); Distinguished Speaker in the field of digital image processing for the Computer Society of India (2015). He served as Distinguished IEEE Lecturer in IEEE India's council for Bombay section. He is the recipient of more than 12 international and national awards, such as the TCS Award 2014, the Rajaram Bapu Patil ISTE National Award 2013, the Engineer of the Year Award 2011, etc. He has delivered more than 50 keynote/invited talks and chaired many technical sessions in international conferences across the world, for example, in Singapore, Myanmar, Sri Lanka, the USA, Italy, and India. His special session on 'Deep Learning in Biometrics' was included in the IEEE International Conference on Image Processing, 2017. He is also a member of many national professional bodies such as the ISTE, CSI, ISCA, and IEI. He is a member of various committees of the University and has been Vice President of the Computer Society of India for the Bhilai chapter for two consecutive years. He is a consultant for various skills development initiatives of NSDC, Government of India. He is a regular referee of project grants under the DST–EMR scheme and several other schemes of the Government of India. He received important consultancy support in the forms of grants and travel support. He has been an expert member of the professor promotion committee of the German–Jordanian University, Jordan and the Project Proposal Evaluation Committee for UK–Israel research grants. Dr Sinha has supervised eight (08) PhD scholars, 15 M. Tech. scholars and is supervising one more PhD scholar. His research interests include biometrics, cognitive science, medical image processing, computer vision, OBE, and ICT tools for developing employability skills.

List of contributors

Ali Ahmadinia
Department of Computer Science and Information Systems
California State University San Marcos
USA

Payal Bansal
Poornima College of Engineering
Jaipur
Rajasthan
India

Aness Banu
Sreenidhi Institute of Science and Technology Yamnampet
Ghatkesar
Hyderabad
India

Maria Bency
Department of Electronics and Telecommunication
RV College of Engineering
Bengaluru
India

Abhishek Bharti
Malaviya National Institute of Technology
Jaipur
India

Lakshmi Sirisha Bollimuntha
ECE Department
V R Siddhartha Engineering College
Kanuru
Andhra Pradesh
India

Rahul Chaurasia
CSE Department
Indian Institute of Technology Indore
Indore
India

Shrikant Chavate
Department of Electronics and Telecommunication Engineering
School of Engineering and Technology
G H Raisoni University
Amravati
India

Surbhi Chhabra
The LNM Institute of Information Technology
Jaipur
Rajasthan
India

Abhishek Choubey
Sreenidhi Institute of Science and Technology Yamnampet
Ghatkesar
Hyderabad
India

Shruti Bhargava Choubey
Sreenidhi Institute of Science and Technology Yamnampet
Ghatkesar
Hyderabad
India

Balwinder Singh Dhaliwal
National Institute of Technical Teachers Training and Research
Chandigarh
India

Aniket Dhole
Department of Computer Engineering
Pimpri Chinchwad College of Engineering
Pune
India

Mohit Gandhi
Department of Computer Engineering
Pimpri Chinchwad College of Engineering
Pune
India

Sonal Gore
Department of Computer Engineering
Pimpri Chinchwad College of Engineering
Pune
India

Kamlesh Gupta
Department of IT
Rustamji Institute of Technology
Gwalior
India

Amardeep Kumar
Malaviya National Institute of Technology
Jaipur
India

Shrishail Kumbhar
Department of Computer Engineering
Pimpri Chinchwad College of Engineering
Pune
India

Kusum Lata
The LNM Institute of Information Technology
Jaipur
Rajasthan
India

Lavanya Maddisetti
Vardhaman College of Engineering
Shamshabad
Telangana
India

Ramasamy Mariappan
ECE department
V R Siddhartha Engineering College
Kanuru
Andhra Pradesh
India

Venkatadri Marriboyina
Dept of Computer Science and Engineering
Amity University Madhya Pradesh
Maharajpura
India

Ravi Mishra
Department of Electronics and Telecommunication Engineering
School of Engineering and Technology
G H Raisoni University
Amravati
India

Ipseeta Nanda
Gopal Narayan Singh University Jamuhar
Sasaram
Bihar
India

Sarah Suhasini Pallikonda
ECE department
V R Siddhartha Engineering College
Kanuru
India

Pavanalaxmi
Visvesvaraya Technological University
Belagavi
India

Lochana Perera
Malaviya National Institute of Technology
Jaipur
India

Shwetha R Prasanna
RV College of Engineering
Department of Electronics and Telecommunication
Bengaluru
India

B S Premananda
Department of Electronics and Telecommunication
RV College of Engineering
Bengaluru
India

Eshaan Rathi
The LNM Institute of Information Technology
Jaipur
Rajasthan
India

J V R Ravindra
Vardhaman College of Engineering
Shamshabad
Telangana
India

Roopashree
Visvesvaraya Technological University
Belagavi
India

Sandeep Saini
The LNM Institute of Information Technology
Jaipur
Rajasthan
India

Pallavi Sapkale
RAIT Nerul
India

Ranjan Senapati
VNR Vignana Jyothi Institute of Engineering and Technology
Hyderabad
Telangana
India

Anirban Sengupta
CSE Department
Indian Institute of Technology Indore
India

Abhishek Sharma
The LNM Institute of Information Technology
Jaipur
Rajasthan
India

Alpana Sharma
Dept of Computer Science and Engineering
Amity University Madhya Pradesh
Maharajpura
India

G P Shreya
Department of Electronics and Telecommunication
RV College of Engineering
Bengaluru
India

Harsh Singhal
Department of Computer Engineering
Pimpri Chinchwad College of Engineering
Pune
India

G R Sinha
Department of Electronics and Communication Engineering
Myanmar Institute of Information Technology
Mandalay
Myanmar

Devendra Kumar Somwanshi
Poornima college of Engineering
Jaipur Rajasthan
India

M Vardhana
NXP Semiconductor
Bangalore
India

Manonita Verma
Malaviya National Institute of Technology
Jaipur
India

Menka Yadav
Malaviya National Institute of Technology
Jaipur
India

Kyle Zampell
Department of Computer Science and Information Systems
California State University San Marcos
USA

Chapter 1

An FPGA implementation of the RSA algorithm using VHDL and a Xilinx system generator for image applications

Sandeep Saini, Kusum Lata, Abhishek Sharma and G R Sinha

Since digital data can be easily copied and manipulated, cryptography is used to secure our data. Whether it is multimedia security or applications related to banking and financial privacy, cryptography algorithms are necessary for protection. Symmetric-key cryptography algorithms such as the Advanced Encryption Standard (AES) require the sender and the receiver to use a shared secret key. However, the whole motivation for security fails if the shared key is not secure enough. Asymmetric algorithms, such as Rivest–Shamir–Adleman (RSA), do not require key security, but they are not feasible if the data are large. However, they can be used for securing small data, for example, images for a multimedia application. Their security relies on the difficulty of the integer factorization problem. The security offered by RSA is due to the existence of huge numbers which are even impossible for supercomputers to factorize in a realistic time. Therefore, this cryptography algorithm is put to good use in various applications to protect electronic data. However, public-key algorithms are computationally intensive. This chapter presents a field-programmable gate array (FPGA) implementation of a 1024-bit RSA cryptosystem on the Virtex-5 XC5VLX2110T-2FF1738 FPGA board. Finite State Machine (FSM) architectures are proposed for key generation, encryption, and decryption. For encryption, the accurate and efficient computation of modular exponentiation is required for 1024 bits; the repeated modular multiplication technique is used for this. For decryption, the L-R binary approach is used, which deploys a modular multiplication module. In our design, efficiency is achieved in terms of throughput/area ratio, whose value is 6.36 megabits per second (Mbps)/slice for encryption and 1.66 Mbps/slice for decryption, which is better than existing

doi:10.1088/978-0-7503-3919-3ch1

implementations described in the literature for the same key size. Furthermore, the encryption and decryption of a gray image are demonstrated with the help of the Xilinx System Generator (XSG) model. Furthermore, to ensure efficient encryption of the shared key by RSA, histograms of the key images before and after encryption are generated and analyzed for encryption strength.

1.1 Introduction

Security techniques involve a large variety of algorithms that are implemented in many applications that require secure communications. The data to be secured can be in the form of plain text, image, audio, or video. The applications that use these security algorithms have application-specific requirements. For instance, mobile handheld devices, smart cards, WSNs, etc. have constrained memory or power. Furthermore, applications such as high-speed network servers require high speeds regardless of the memory or power. Moreover, there must be a trade-off between the implementation cost and the level of security offered by the implementation. A larger key size provides a greater level of security, but it also results in larger memory use and increased complexity, hence greater cost.

If the information to be secured is large, cryptography algorithms such as AES, IDEA, Twofish, Blowfish, etc. come into play. As stated earlier, if it has to be implemented in hardware, the choice of the algorithm used to secure data depends on the applications and constraints. The algorithms mentioned above are faster to implement because they use a secret (key) that is shared between the sender and the receiver. But this causes a major security concern—the security of the shared secret (key). The algorithms that use a shared key are termed symmetric cryptography algorithms, since the shared secret is common to both the sender and the receiver, and is used for encryption and decryption at the respective ends.

Due to the dual security (of the message and the key) needed in symmetric algorithms, asymmetric cryptography algorithms such as RSA were developed. These algorithms are called asymmetric because the key is no longer shared by the sender and the receiver. This means that the key used for encryption is different from the key used for decryption. Hence, unlike the symmetric algorithms, there is no security concern related to sharing the key. So, we can use asymmetric-key algorithms, such as RSA, to resolve the issue of key security in symmetric-key algorithms. Apart from this, since asymmetric-key algorithms are computationally intensive, they tend to be slower and more complex; hence, they are more suitable for small message sizes. This makes them suitable for image applications. RSA's security is based on the difficulty of factorizing large integers [3, 4].

The strength of the security lies in the length of its keys. The larger the key, the more secure it is. However, the level of security needed may vary from application to application. Hence, the key length can be tailored accordingly. This requires ease of reconfiguration of the key length, which is a remarkable feature of FPGAs. FPGAs have programmable logic cells that can be programmed according to the user and generate the desired functionality. Since they are not hardwired to a fixed architecture, they can be reconfigured [5]. They are coded in a hardware description language (HDL) and usually have various process paths running concurrently. This parallel processing feature makes them speedier than conventional processors.

1.2 Literature review

1.2.1 Introduction to cryptographic algorithms

Various cryptographic algorithms exist, which vary in the levels of security that they offer and the key mechanisms used. They are mainly classified into the symmetric and asymmetric categories. These algorithms are further described in the following subsections.

1.2.2 Types of cryptographic algorithm

1.2.2.1 Private-key cryptosystems

Private-key cryptosystems, or symmetric cryptosystems as they are called, use a key that is shared between the sender and the receiver. That shared key must be kept secret (private), as shown in figure 1.1.

Examples include cryptography algorithms such as the Data Encryption Standard (DES), the AES, Blowfish, the International Data Encryption Algorithm (IDEA), etc. Symmetric-key cryptosystems are faster, but they have various drawbacks, such as requiring additional security for key exchange. Apart from the key security, these systems only provide confidentiality. There is no way to check the authenticity, because digital signatures are not possible with symmetric-key algorithms such as AES [6]. Moreover, for a network that is not restricted to one-to-one communication, which has a number of users n, $n(n-1)/2$ different keys will have to be generated. Managing so many keys is a cumbersome task, especially if n is large.

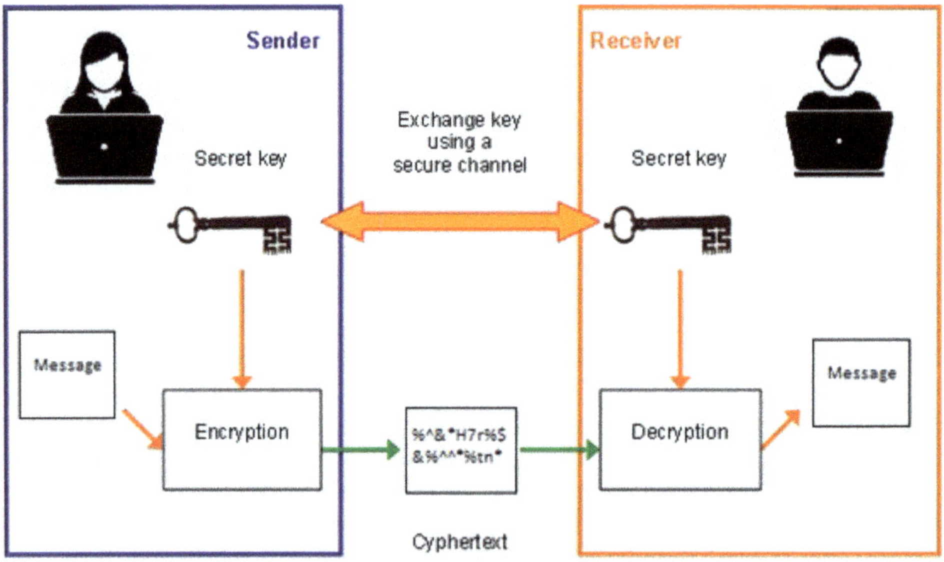

Figure 1.1. Symmetric-key cryptosystem.

1.2.2.2 Public-key cryptosystems

Unlike private-key cryptosystems, public-key cryptosystems do not have to meet the requirement of keeping the shared attributes secret. This is because of the concept of a public and a private key for each user involved in communication. Each user has their own set of two keys: one of them is public (known to everyone); the other is private and not known to anyone except the user.

Therefore, the sender has two keys, public and private, and the receiver has its own set of public and private keys. Since the public key is known to everyone, the receiver's public key is known to the sender and vice-versa. For encryption, the sender encrypts the message using the receiver's public key. The ciphertext generated is then decrypted by the receiver using their private key, as shown in figure 1.2. Since nobody except the receiver has the receiver's private key, only the receiver can decrypt the ciphertext.

Interestingly, not even the sender can decrypt the ciphertext, because the cipher can only be decrypted using the receiver's private key. Also, an important thing to note is that the receiver's private key and the public key are generated such that they are mathematically related to each other. The calculation of the private key from the public key is infeasible. The key generation is based a trapdoor function, which is easy to compute in one direction but hard to compute in the reverse direction [1]. The two most famous trapdoor functions are based on the integer factorization problem used in RSA and the discrete logarithmic problem used in elliptical curve cryptography (ECC).

1.2.3 Important terminology

Some common but important terminology used in cryptography is briefly described below:

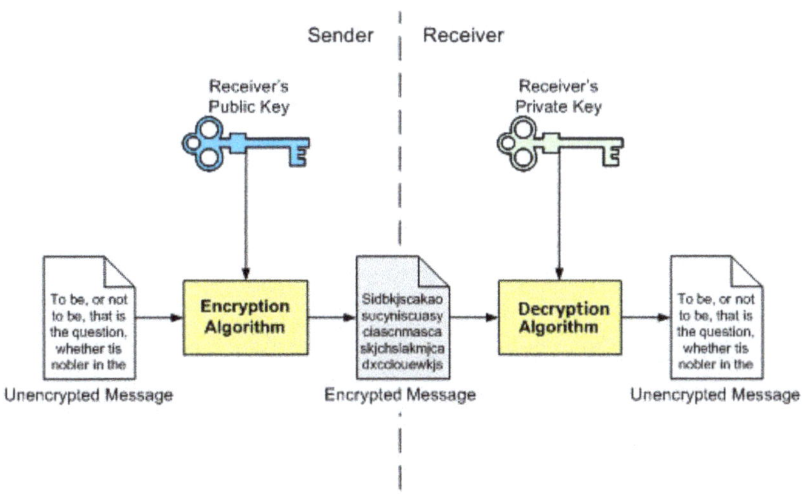

Figure 1.2. Asymmetric-key cryptosystem.

Table 1.1. Comparison of different cryptography algorithms.

Algorithm	Year	Strategy	Advantages	Disadvantages
DES	1975	Fiestel function	Obsolete	The key size is too small; hence, DES is considered insecure now.
3-DES	1998	The DES algorithm is applied to each block three times.	The key length is thrice that of DES, hence it is more secure than DES.	Vulnerable to differential cryptanalysis attacks [12]
RC2	1996	16 mixings and 2 mashings	It can be used in all the modes that DES can be used in.	Vulnerable to differential and brute-force attacks [12]
Blowfish	1993	Fiestel network [13]	One of the fastest block ciphers	Vulnerable to differential and brute-force attacks [12]
AES	1976	Substitution and permutation network	Simple design and fast computation. Among symmetric algorithms, this is the most secure.	Side-channel attacks and cache collision attacks can recover the whole key [14].
RSA	1977	Integer factorization problem	Even the fastest algorithm to break RSA-1024 takes a sub-exponential time	For security for a long period, a modulus size greater than 1024 bits should be used [15, 16].
Rabin	1980	Integer factorization problem	Encryption is fast because of single modular squaring [17]	Decryption is slow. Vulnerable to RSA attacks [2, 17].
ECC	1985	Discrete logarithmic problem	Provides the same level of security as RSA, but with a lower key size	The intermediate value of the cipher is very long, i.e. comparable to RSA's modulus. Pollard's parallelizable rho method can attack discrete logarithmic problems [18, 19]
ElGamal	1985	Diffie–Hellman and a discrete logarithmic problem	Encryption is by randomization	Two modular exponentiations in encryption slow down the encryption process [20] The ciphertext is a factor of two and is prone to a chosen ciphertext attack

Table 1.2. Comparison of different implementations of RSA and its modules.

Algorithm	Year	Approach	Resources	Results
Sharma [21]	2016	Montgomery algorithm for encryption	Xilinx Virtex-5XC5V LX50–3FF676	Slices utilised = 8012
Thobbi [22]	2015	Multiplication performed using Booth's algorithm	XilinxSpartan3X C3s400fg320–4	LUT utilised = 1134, frequency = 67.696 MHz, power = 0.086 W
Shehata [23]	2014	1024-bit encryption performed using a square-and-multiply technique	Xilinx Virtex5xc5vtx 240t-2ff175	Slices utilized = 28 350, max freq. = 36.290 MHz
Song Bo [24]	2011	Modular exponentiation using Montgomery modular multiplication	Xilinx Virtex-6 family FPGA XC6VLX240T-1	RSA module runs in 277.26 ms.
Nakano [25]	2009	Modular exponentiation using redundant 64 K number system	Xilinx Virtex-II Pro-family FPGA XC2VP30–6	1024-bit modulo exponentiation in less than 2.521 ms
Ghoreishi [26]	2009	Based on modified Booth's technique and Montgomery multiplication	Xilinx Virtex-II and XC4000 series FPGA	RSA operation in less than 15 and 50 ms on Xilinx Virtex-II and XC4000 series FPGA, respectively

- **Encryption**: Encryption is performed on a message to convert it into a cipher (an unreadable form) for the sake of confidentiality. This is done by the sender, who uses a key. Encryption ensures that the message can only be decrypted by the intended user(s).
- **Decryption**: Once the encrypted message is received by the receiver, it must be decrypted. Like encryption, the decryption process uses an algorithm and a public or a private key depending on how the message has been encrypted.
- **Confidentiality**: Confidentiality is the protection of information from access by unauthorized parties. Since information like bank account details, government documents, etc. is nowadays sensitive, it needs to be secured from eavesdroppers.
- **Authenticity**: Since information is communicated electronically, authentication is used to ensure that the message has been sent by the authorized sender only and not by some malicious party. This is done by confirming the receiver's identity using a digital signature.
- **Integrity**: The integrity of the message is preserved by ensuring that the message was not changed or tampered with after the sender sent it and before the receiver received it. To maintain the integrity of the messaging, hashing techniques are commonly applied.
- **Non-repudiation**: There also may be instances when the sender denies that he has sent any message, even when he has sent it. Non-repudiation then comes into play by legally proving that the sender sent the message. This proof can be verified without the consent of the said sender.

1.2.4 Comparison of different cryptographic algorithms

See table 1.1.

1.2.5 Comparison of the implementations of different cryptographic algorithms

See table 1.2.

1.3 Mathematical formulation

Consider integers x and y which are congruent modulo n, which is represented as $x \equiv y \bmod n$. This means that $x - y$ is divisible by n, or, in other words, when x is divided by n, it gives n as the remainder. Certain properties [7] of unity must be kept in mind and are used to reduce large computations. It may be noted that RSA involves large computations.

1. If $x \equiv y \bmod n$, then $y \equiv x \bmod n$. Also, transitive properties follow, i.e. if $x \equiv y \bmod n$ and $y \equiv z \bmod n$, then $x \equiv z \bmod n$.
2. If $x \equiv y \bmod n$ and $u \equiv v \bmod n$, then $(x + u) \equiv (y + v) \bmod n$.
3. If c is an integer and $x \equiv y \bmod n$, then $cx \equiv cy \bmod n$.
4. If $x \equiv y \bmod n$ and $u \equiv v \bmod n$ then $xu \equiv yv \bmod n$.
5. If k is a positive integer and $x \equiv y \bmod n$, then $xk \equiv yk \bmod n$.
6. $xu \equiv yv \bmod n$ implies that $x \equiv y \bmod n/\gcd(u, n)$.

1.3.1 Bézout's theorem

Consider two non-zero integers, r_0 and r_1. Let the greatest common divisor (GCD) of r_0 and r_1 be denoted by g, such that d = GCD(r_0, r_1). According to Bézout's theorem, there exist integers s and t such that

$$sr_0 + tr_1 = g. \quad (1.1)$$

The GCD can now also be denoted by GCD(r_0, $R - 1$) = (g, s, t), and equation (1.1) is called Bézout's form [7]. The integers s and t are called Bézout's coefficients. They can be found using the extended Euclidean algorithm, which is discussed further in this chapter. Also, many other theorems, such as Euclid's lemma or the Chinese remainder theorem (CRT), result from Bézout's form.

1.3.2 The Euclidean algorithm

The GCD can be found easily using prime factorization. However, for large numbers, this method becomes cumbersome. Therefore, the Euclidean algorithm (a reductive method) is used, which finds the GCD of two numbers, say r_0 and r_1, without actually factoring.

The Euclidean algorithm states that gcd(r_0, r_1) = gcd(r_0 mod r_1, r_1). This can be explained with the help of an example, as shown below.

Consider r_0 = 863 and r_1 = 197. GCD(863, 197) is found using the Euclidean algorithm as shown below:

GCD(863, 197) = GCD(863 mod 197, 197)
= GCD(197, 75)
= GCD(197 mod 75, 75)
= GCD(47, 75)
= GCD(75 mod 47, 47)
= GCD(28, 47)
= GCD(47 mod 28, 28)
= GCD(19, 28)
= GCD(28 mod 19, 19)
= GCD(9, 19)
= GCD(19 mod 9, 9)
= GCD(1, 9)
= 1

We can observe that we have reduced the problem of calculating the GCDs of large numbers to the calculation of the GCDs of small numbers, which is computationally feasible.

1.3.3 The extended Euclidean algorithm

The GCD of r_0 and r_1 can be written as $sr_0 + tr_1$. The extended Euclidean algorithm makes use of the Euclidean algorithm to find the coefficients s and t. The need to find these coefficients arises in the RSA algorithm while computing the private key. Hence, this algorithm plays a very important role.

This algorithm makes use of the new remainder generated at every iteration of the Euclidean algorithm. Let the new remainder for the ith iteration be denoted by r_i. The value of r_i is computed at every iteration and is represented as shown in equation (1.2).

$$r_i = s_i r_0 + t_i r_1. \tag{1.2}$$

The linear combination shown in equation (1.2) is constructed with the help of similar linear combinations from previous iterations [8]. The two previous iterations can be represented using the construction shown in equations (1.3) and (1.4).

$$r_{i-2} = s_{i-2} r_0 + t_{i-2} r_1 \tag{1.3}$$

$$r_{i-1} = s_{i-1} r_0 + t_{i-1} r_1. \tag{1.4}$$

The new remainder at the ith iteration is calculated from the quotient q_{i-1} and the previous remainders, r_{i-2} and r_{i-1}, as follows:

$$r_{i-2} = q_{i-1} r_{i-1} + r_i. \tag{1.5}$$

Equation (1.5) can be rewritten as follows:

$$r_i = r_{i-2} - q_{i-1} \cdot r_i - 1. \tag{1.6}$$

To represent r_i as shown in (1.2), we substitute the values of r_{i-2} in equation (1.3) and r_{i-1} in equation (1.4) into equation (1.6). On substituting and rearranging the terms, we get equation (1.7), which is further simplified as equation (1.8).

$$r_i = [s_{i-2} - q_{i-1} s_{i-1}] r_0 + [t_{i-2} - q_{i-1} t_{i-1}] r_1 \tag{1.7}$$

$$r_i = [s_i] r_0 + [t_i] r_1. \tag{1.8}$$

Therefore, from equation (1.8), we can compute the values of s_i and t_i. However, it should be observed that we need their values from the previous iterations for the calculations of s_i and t_i. Therefore, the initial values are taken to be $s_0 = 1$, $s_1 = 0$, $t_0 = 0$, and $t_1 = 1$ [8].

It may be noted that for the computation of the modular inverse, it is sufficient to know the value of t_i.

1.3.4 Euler's totient function and Euler's theorem

Euler's totient function, denoted by $\Phi(n)$, is the number of integers represented by the variable a, in the range $1 \leqslant a \leqslant n$, which are relatively prime to n. A number a is relatively prime to n if $\gcd(a, n) = 1$. Below are some important properties of Euler's totient function, $\Phi(n)$.

1. If p is prime, then $\Phi(n) = p - 1$.
2. $\Phi(n)$ is multiplicative. If $\gcd(m, n) = 1$, then $\Phi(mn) = \Phi(m)\Phi(n)$.

For small numbers, $\Phi(n)$ can be calculated by listing all the numbers from 1 to n and then counting how many are relatively prime to n. This naive approach, however, is feasible only if n is small. If n is large, the following theorem is used if the factorization of n is known [8]:

Let the canonical factorization representation of pi be distinct prime numbers, and e_i are positive integers. Then $\Phi(n)$ can be found as:

$$\Phi\left(n\right) = \Pi_{i=1}^{k} p_i^{e_i} - p_i^{e_i - 1}. \tag{1.2}$$

Euler's totient function is used to establish Euler's theorem, which further establishes the concept of a secure cryptosystem such as RSA. The theorem states that if $\gcd(a, m) = 1$, then $a\Phi(m) \equiv 1 \bmod m$.

1.3.5 Modular exponentiation techniques

Techniques for the computation of modular exponentiation are needed due to cryptography algorithms such as RSA, which deal with a large number of bits in computations. Even if the number of bits is not large, calculating modular exponentiation in the conventional way is inefficient in terms of the time taken for the computation and the number of computations needed. Therefore, calculating modular exponentiation significantly slows down the system and hence affects performance.

1.3.5.1 The repeated modular multiplication approach

The computation of $z = xy(\text{mod } n)$ is nothing but the repeated modular product. The algorithm used to find modular exponentiation is as follows:

1. Set $z = 1$, $y_0 = 0$.
2. Increase y_0 by 1.
3. Set $z = (yz) \bmod n$.
4. If $y_0 < y$, goto step 2. Else, $z = xy(\text{mod } n)$.

This can be explained with the help of an example, as follows:

Let $x = 85$, $y = 5$ and $n = 497$.

1. for $y_0 = 1$, $z_0 = (1 * 85) \bmod 497 = 85 \bmod 497 = 85$
2. for $y_0 = 2$, $z_0 = (85 * 85) \bmod 497 = 7225 \bmod 497 = 267$
3. for $y_0 = 3$, $z_0 = (267 * 85) \bmod 497 = 22\,695 \bmod 497 = 330$
4. for $y_0 = 4$, $z_0 = (330 * 85) \bmod 497 = 28\,050 \bmod 497 = 218$
5. for $y_0 = 5$, $z_0 = (218 * 85) \bmod 497 = 18\,530 \bmod 497 = 141$

This algorithm, explained by the above example, uses the fact that for two given integers x and y, equations (1.9) and (1.10) are equivalent

$$z_0 \bmod n = (x \cdot y) \bmod n \tag{1.9}$$

$$z_0 \bmod n = [(x \bmod n)(y \bmod n)] \bmod n. \tag{1.10}$$

1.3.5.2 The square-and-multiply technique

This method is also called the binary method. It has two variants, the left-to-right (L-R) binary method and the right-to-left (R-L) binary method [24], [12], [25].

The L-R binary method

Inputs: M, e, n

Output: $C := Me \bmod n$

1. if $e_{h-1} = 1$, then $C := M$ else $C := 1$
2. for $i = h - 2$ down to 0
   (a) $C := C \cdot C \bmod n$
   (b) if $e_i = 1$, then $C := C \cdot M \bmod n$
3. return C

In the left-to-right binary method, the bits of the exponent are scanned from left to right, i.e. from the most significant bit to the least significant bit. The output value is initialized to 1 initially, and modular squaring is done for each bit of the exponent scanned. If the bit is 1, a modular multiplication is performed.

The L-R method requires $e-1$ squaring operations and about an average number of multiplications.

The R-L binary method

Inputs: M, e, n

Output: $C := Me \bmod n$

1. $C := 1$; $P := M$
2. for $i = 0$ to $h - 2$
   (a) if $e_i = 1$, then $C := C \cdot P \bmod n$
   (b) $P := P \cdot P \bmod n$
3. if $e_{h-1} = 1$, then $C := C \cdot P \bmod n$
4. return C

In the right-to-left binary method, the bits of the exponent are scanned from right to left, i.e. from the least significant bit to the most significant bit. This method uses an intermediate variable P to store the values of the intermediate powers. The output value is initialized to 1 initially, and modular squaring is done for each bit of the exponent scanned. If the bit is 1, a modular multiplication is performed. Apart from the scanning direction and the use of an intermediate variable to store the powers, the procedure remains the same as in the L-R binary technique.

The L-R method and R-L method require $e - 1$ squaring operations and about an average number of multiplications [9]. However, the R-L method requires one more register to store the intermediate powers. It may be noted that the steps in the R-L method can be performed in parallel when they are implemented, because they are independent of each other [10].

1.3.6 Chinese remainder theorem

For modular exponentiation, where, apart from the modulus, the exponent is also large, the CRT is used. RSA decryption is one such case, because in order to maintain the security level, the length of the private-key exponent d cannot be compromised.

However, it may be noted that the decryption end of RSA that calculates $xd \bmod n$ has access to the two prime numbers from which the keys are generated.

So the basic idea of the CRT is that instead of performing modular exponentiation on one long modulus, we do the same on the two prime numbers, p and q, which are almost half the length of the modulus n. This is a kind of transformation domain arithmetic. It is divided into three main parts: transforming into the CRT domain, computing in the CRT domain, and transforming the result back to the real domain.

1.4 RSA system

The RSA algorithm is based on the problem of the integer factorization of large numbers. It utilizes two keys, one of which is public and the other of which is private. Since the public key is known to everyone, the receiver's public key is known to the sender and vice-versa. For encryption, the sender encrypts the message using the receiver's public key. The ciphertext generated is then decrypted by the receiver using their private key. The RSA algorithm is divided into three parts: key generation, encryption, and decryption.

1.4.1 Key generation

The steps for key generation are as follows:

1. Two prime numbers, p and q, of roughly equal lengths are generated randomly.
2. Calculate $n = p * q$
3. Compute Euler's totient function, $\Phi(n)$ as: $\Phi(n) = (p - 1) * (q - 1)$
4. Choose an integer e such that it satisfies the following two conditions:
   (a) $1 < e < \Phi(n)$
   (b) $\mathrm{GCD}(e, \Phi(n)) = 1$
5. Calculate d such that it is the multiplicative inverse of e, i.e. $d \equiv e - 1(\bmod\ \Phi(n))$. This means
   $e \cdot d \equiv 1(\bmod\ \Phi(n))$.
6. (e, n) and (d, n) are the public key and the private key (of the receiver), respectively.

1.4.2 Encryption

The steps for encryption are as follows:

1. The sender first obtains the receiver's public key (n, e).
2. The message to be encrypted is represented as an integer m, such that $m > 0$ and m lies in the interval $[0, n - 1]$.
3. The sender then computes cipher c, which is $me(\bmod\ n)$.
4. The cipher c is then sent to the receiver for decryption.

1.4.3 Decryption

The steps for decryption are as follows:

1. The receiver, upon generating its public and private keys, will receive the cipher from the sender.
2. The cipher will be decrypted as $m = cd(\bmod\ n)$.

The steps involved in encryption and decryption may seem easy to compute at first. The challenge is to implement modular exponentiation for such large numbers with accuracy and efficiency in hardware.

1.5 System implementation

1.5.1 Key generation

1.5.1.1 Random 512-bit prime number generation

A 512-bit-long odd random number is generated and tested for primality using the Rabin Miller test [11]. The random number is generated by designing a pseudo-random number generator based on an XOR operation on a 512-bit-long shift register. The internal architecture of the random number generator is shown in figure 1.3.

Figure 1.4 shows the generation and verification of 512-bit random numbers. The value ACK = 1 confirms that the number is prime. The simulation result for the same is shown in figure 1.5. Likewise, another prime number is generated, so we have two 512-bit prime numbers, p and q.

*1.5.1.2 Generation of the public key (*n, e*) and the private key (*d, e*)*

In this step, $n = p * q$ and $\Phi(n) = (p - 1) * (q - 1)$ are calculated.

For the calculation of e, the GCD is found using the Euclidean algorithm, as shown in figure 1.6. To calculate the public key, the state Inc._PubKey increments the variable PubKey by 1 and the state Check_Large checks for the larger of Phi and

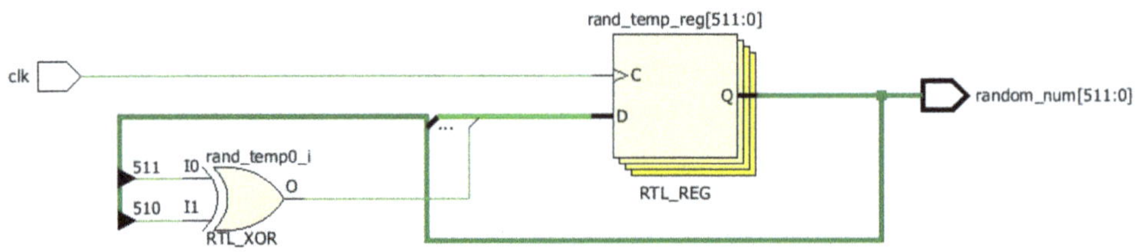

Figure 1.3. Internal architecture of a 512-bit pseudo-random number generator using a 512-bit shift register.

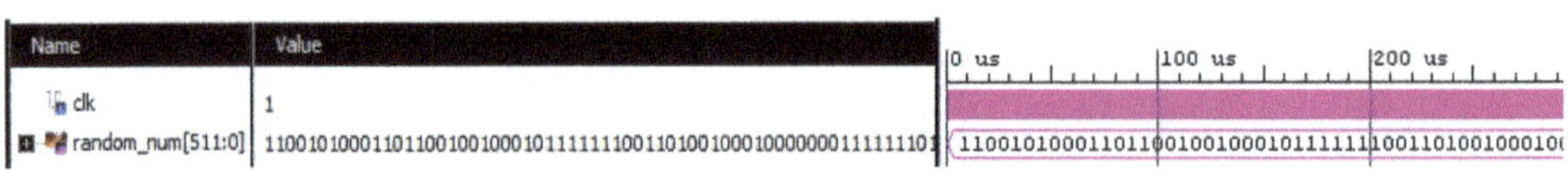

Figure 1.4. Simulation showing the generation of random 512-bit number which is a prime number.

Figure 1.5. Simulation showing confirmation of a random 512-bit number that is a prime number.

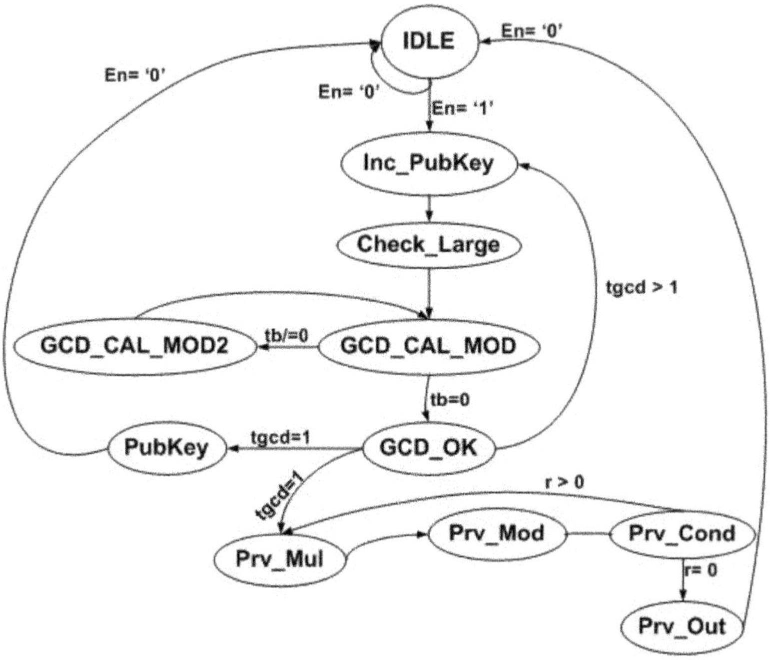

Figure 1.6. State flow in public and private key generation.

PubKey. Accordingly, the GCD is calculated by states GCD_CAL_MOD and GCD_CAL_MOD2 using the Euclidean algorithm by repetitive subtraction. The GCD is found when the smaller of Phi and PubKey is reduced to zero. The state GCD checks whether the GCD found, i.e. tgcd, is equal to 1. If it is not equal to 1, the current state is set to the Inc._PubKey state. This process continues until the GCD is found to be 1. Once the GCD is 1, the corresponding value of PubKey is the public-key exponent, *e*.

Now, to compute the private key, the condition $e \cdot d \equiv 1(\text{mod } \Phi(n))$ can be rewritten as $d = (1 + (k * \Phi(n)))/e$, for the smallest value of k, where $k > 0$. The states Prv_Mul and Prv_Mod compute the value of the remainder when $(1 + (k * \Phi(n)))$ is divided by e, assuming $k = 1$ initially. The state Prv_Cond checks whether the remainder is 0. If not, k is incremented by 1, and the current state is set to the Prv_Mul state.

This goes on until $(1 + (k * \Phi(n)))$ is completely divisible by e, for some value of k. The corresponding value of d computed is the private key exponent, d. Once the public-key exponent, e and the private key exponent, d, have been computed, encryption and decryption can be done, respectively.

Thus, we get the two prime numbers, p and q, as shown in table 1.3.

1.5.2 Encryption

The cipher can be calculated as $c = me(\text{mod } n)$. Here, m is the plain text or message to be encrypted, e and n are the public-key exponent and the modulus, respectively, generated by the key-generation module. Since the value of n is very large (1024 bits

Table 1.3. Various RSA parameters and their generated values.

Parameter generated	Value
p	1100101000110110010010001011111110011010010000110010101111101101101010000010101000111000101010001010101110001110011100001000010001101111001000011000011111011101011001110010100011000111100110110100110011100101111111101010000010100110010000000111100001011000001101110010010110001110110011011001001010010000000111010011101011111010000001101111011101000001100010010101110100101111100111100001000010001000101010010001001111100111001101101001111100011010010011111010000000001101111 1110101110101000110011001011000110110 01
q	10010101110000011001111001011011011110100010100010110111111100100010001100111001110101101011111 10000111000011010110101101101001011001101010000110110101101111011100011000111100110101101101101011 1110110101100110111110001010110011100010010001001111011001111001001011010101011010010001100010 0100011100000001101000011011100101011010010001100111010010110001110010111011111101111000011 1011001100011010011000110100001111111011110000001101001000100111110101100000100010101101110010110 11010101001001000100011110100001111011 1
n	0011101100100101010001000010110100001111111011001000101011111001011110101010001010101001000100010 0 0111010110011110000110101000011111000101001100011010001110010001110001110010100010011111001 1 1010011111100110011001001011001001101111111001000101000101000101010100100111100001000010101101 0011001001000111111001101110101101010000100010100100011010100111100101000101100000110011100011 0111010101010011111110001111100001010011100101101001110010101110000001011110101010010001000 11 101010101111110111111000100001001010110000001100100000010001011100111100011000110001 01100 0011110110101010001010001111000001110101001000100010110101011101000001100111011101100000 01 0010001100001100110111000001000110110010010110101100111100111000000100111000100111001110000100 1100111011111011011100010001100010011010010111110110001100000101011111110101111010110111 01001 00110011100001000011101100000110011010011001100000011010101101110000110100101100111001101001 1 101110000100000001100011010100110100001001100110000000000010001101111111010011011000010010101111 1
$\Phi(n)$	0011101100100101010001000010110100001111111011001000101011111001011110101010001010011011100100 0 00101111101011111011100000011101101110110001111110110101010111011100001011001111111011100 0010001000000000110110001101110011100000101110010011111101101010100001110101101100110010

(*Continued*)

Table 1.3. (*Continued*)

Parameter generated	Value
	0000001000101110110000011110010110101011110010111010010000101011001111101100100011000100000000 011100111001010100011010101100001000011010111111010010011101011000000011010111001110000110101 011100000011000111000011010011110011000001011000110111011010000010110010000100010001111111 1001110100100100100011111000001101001110110010100100111101101110111100000000000100000 10010111100110001110000100101010111100101110000100011110011111101100100011100010 000001000000011111011100101110110100100001101101100100011010000100100000000000111010110100 1011001110000111110100110100010010101111000101101111101011110001001010110001100011100000101 100001010001111110110100100010000101101010010011101111110001001111111011001101111111110011 1001101001100000010000110111111001000
e	011
d	10011101101110001011010111001101011111111110011000001110101000011111100011011000110011110110101100101 00 1110101001001010110100100100111011010100111100011101001001011001000101010011110101101100000000 0 010010000100100110100000000111101101111111100111000101101000111100110011000000000101110100 1 0000001010000111100011111011101000110000001110011010100100001011101100000000000100110100001 11000010001110010110000010001111111100001101000111010101100111010001001011001110010010101 1 001011110101111000101000100000000111011001111100110101100100001011000001011111111111011111000 01100001011111110101100110111110011000110001010010010011111010101010101101011000011111011 10 110100000011000111010011001001011000010100010101001000010111101100000000010101011111110100 00111110011011010111100111011011001101011011010101010110100011110000111011110101111111000 1 0001011011100101000001111010100011111011000011100101110110100000011101011100001010100100 01 10000010110010001101101111101010010110111111111001100111111111101111011110001000000001011 0 01111111011011

in our case), computations with such a large value are not possible. Therefore, modular exponentiation is needed to calculate the cipher with the help of the repeated modular multiplication technique.

1.5.2.1 Modular multiplication

Modular multiplication means calculating $(A * B) \bmod n$. It can be implemented using various techniques. In our design, the shift-and-add technique was used for the 1024-bit multiplication operation. The multiplicand and the multiplier were k-bit binary numbers, where $k = 1024$. This is shown by the top-level schematic of the modular multiplication module in figure 1.7.

The partial product modulo was also reduced at every step:

1. $P = 0$
2. for $i = 0$ to $k - 1$
   (a) $P = 2P + A \cdot Bk - 1 - i$
   (b) $P = P \bmod n$
3. Return P

The P generated might be greater than the modulus n. Therefore, two subtractions are needed so that P lies in the interval $[0, n - 1]$. The subtractions performed are as follows:

1. If $P - n > 0$, then $P = P - n$
2. If $P - n > 0$, then $P = P - n$

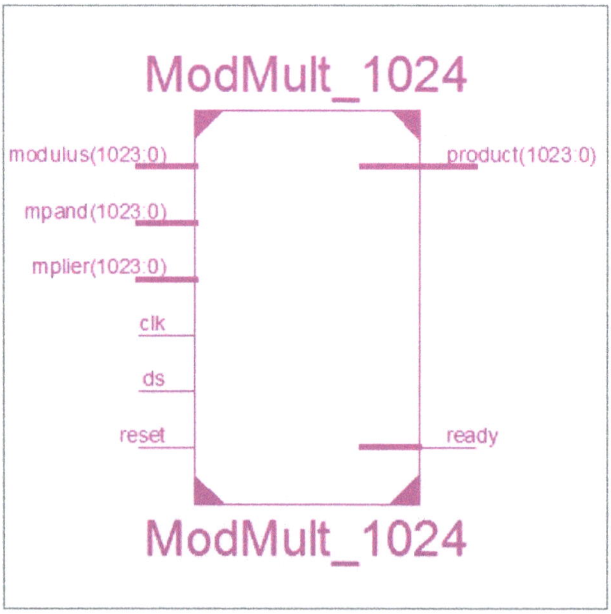

Figure 1.7. Top-level architecture of the modular multiplication module.

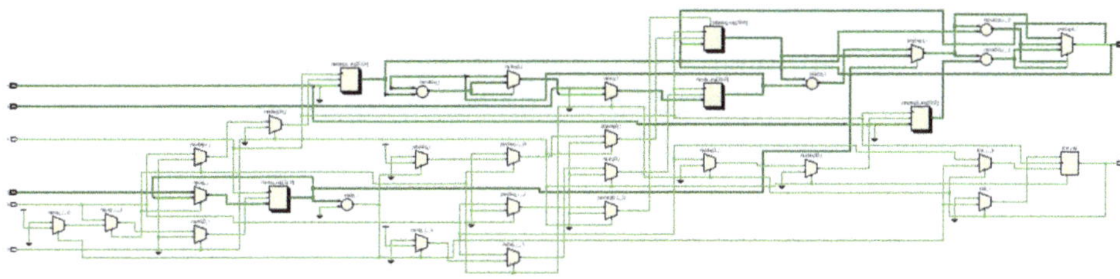

Figure 1.8. Internal architecture of the modular multiplication block.

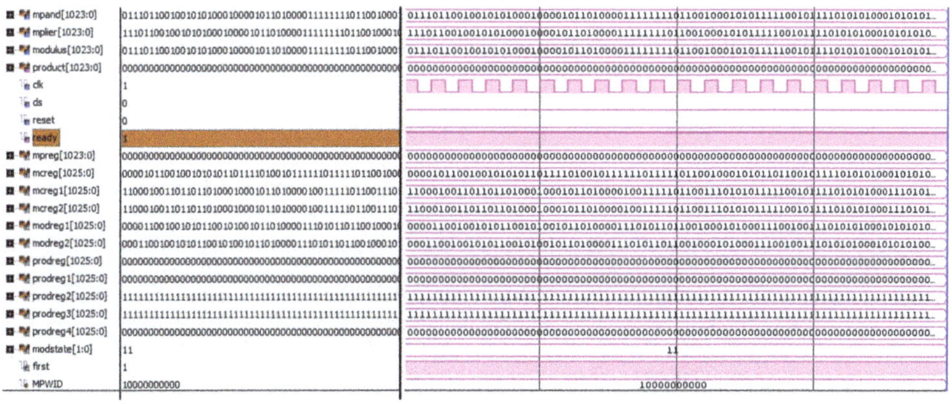

Figure 1.9. Simulated result of 1024-bit modular multiplication.

The same subtraction is done twice because P is incremented twice in the algorithm. In step 2(a), once it is incremented to $2P$, it is further incremented by the addition of $A * Bk - 1 - i$. Figure 1.8 shows the generated internal architecture of the modular multiplication module.

Figure 1.9 shows the results of a simulation of the modular multiplication module generating the value of $(N * 2N) \bmod N$, which is zero.

1.5.2.2 Modular exponentiation

The cipher $c = me(\text{mod } n)$ is nothing but a repeated modular product. The modular product, $(A * B) \bmod n$ is deployed in the repeated modular multiplication technique to perform modular exponentiation. The repeated modular multiplication technique can be explained with help of a short example, as follows:

Let $m = 85$, $e = 5$, and $n = 497$.

- for $e_0 = 1$, $c_0 = (1 * 85) \bmod 497 = 85 \bmod 497 = 85$
- for $e_0 = 2$, $c_0 = (85 * 85) \bmod 497 = 7225 \bmod 497 = 267$
- for $e_0 = 3$, $c_0 = (267 * 85) \bmod 497 = 22\,695 \bmod 497 = 330$
- for $e_0 = 4$, $c_0 = (330 * 85) \bmod 497 = 28\,050 \bmod 497 = 218$
- for $e_0 = 5$, $c_0 = (218 * 85) \bmod 497 = 18\,530 \bmod 497 = 141$

A state flow approach is used to implement the encryption block. The flow of states for encryption is shown in figure 1.10.

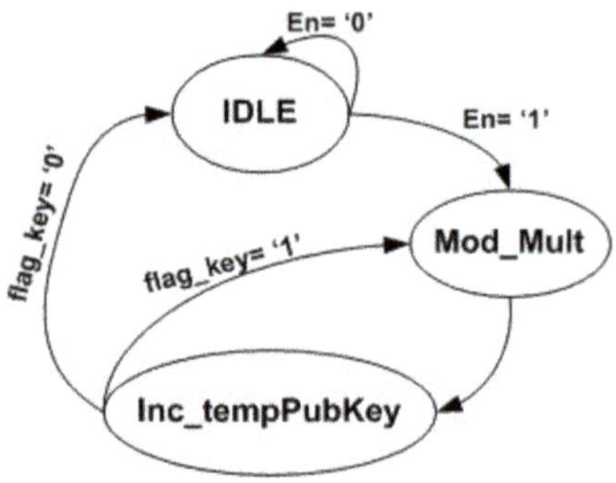

Figure 1.10. States involved in the implementation of the encryption block.

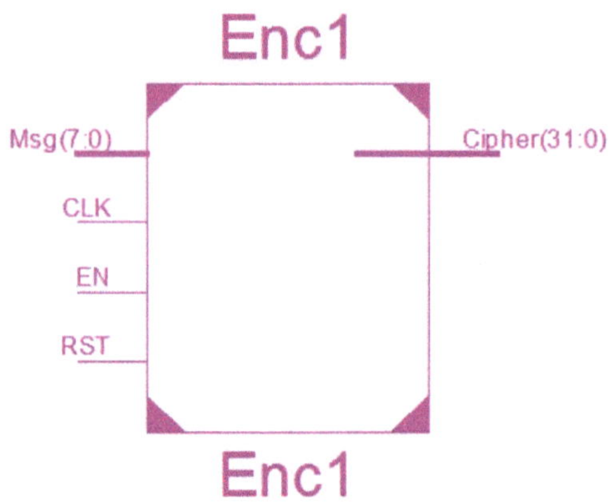

Figure 1.11. Top-level diagram of the encryption block.

The public-key exponent *e* is incremented by one iteration, and the signal flag_key is set to 0 as soon as all the iterations are complete. When flag_key is 0, the encryption is done, and the current state changes to the idle state.

The top-level architecture used for encryption is shown in figure 1.11.

The message to be encrypted is a gray image. Therefore, the maximum value of a pixel is 255. Hence, an 8-bit message length is used as the input of the encryption block. The cipher generated has 32 bits. Figure 1.12 shows the generated internal architecture of the encryption block. The submodule cipher calculation is nothing more than the modular multiplication module. This module is instantiated as P1 and is called multiple times by the main encryption program.

The simulated encryption results are shown in figure 1.13. The message '11111111' is encrypted into a 32-bit cipher, as shown below.

1.5.3 Decryption

Unlike encryption, here we have a large exponent value (the private key, *d*). Therefore, the left-to-right binary approach is used. Assuming *d* is *k* bits long, the approach is as explained below:

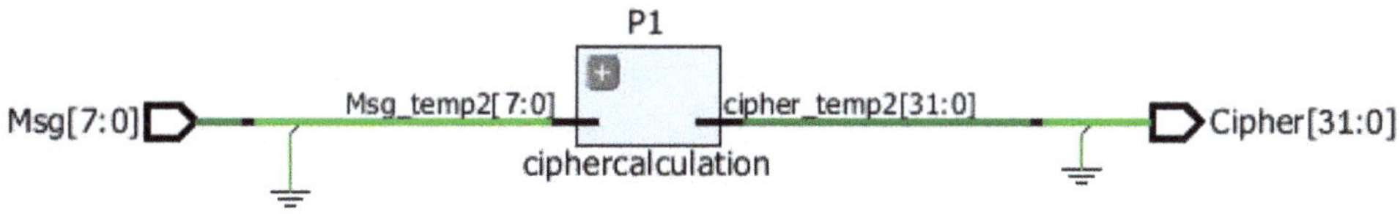

Figure 1.12. Schematic showing that the encryption block has a submodule for cipher calculation.

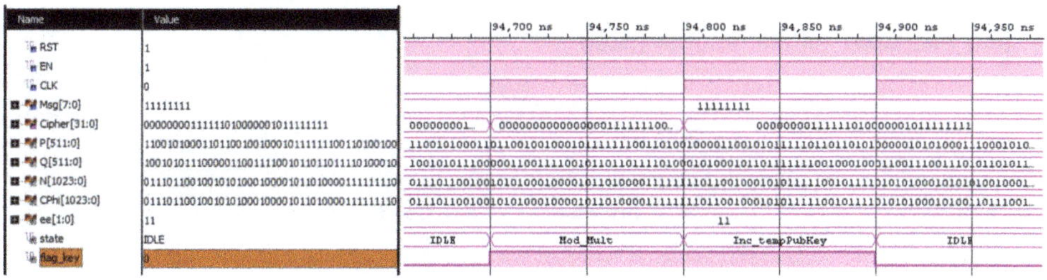

Figure 1.13. Simulated encryption results.

Inputs: c, d, n

Output: *Decmsg* := cd(mod n)

1. if $d_{k-1} = 1$, then *Decmsg* = c else *Decmsg* = 1
2. for $i = k - 2$ down to 0
   (a) *Decmsg* = (*Decmsg* $*$ *Decmsg*) mod n
   (b) if $d_{i=1}$, then *Decmsg* = (*Decmsg* $*$ c) mod n
3. Return *Decmsg*

The value of *Decmsg* in steps 2(a) and 2(b) is calculated by the modular multiplication module. The states used to implement the decryption block are shown in figure 1.14. After the private key exponent d has been calculated, the k bits of d are scanned from left to right. State Initialise initializes *Decmsg* according to the value of the leftmost bit of d, i.e. d_{k-1}. After the initialization, the state Mod_Mult computes (*Decmsg* $*$ *Decmsg*) mod n. The variable temp is updated, depending on whether the current bit of d_{k-1} is 1 or not. Modular multiplication is computed again if $d_{k-1} = 1$. As shown in figure 1.14, this flow of states continues until all the bits of d have been scanned. The value of *Decmsg* after the last bit is scanned is the decrypted message value.

A top-level schematic of the decryption block is shown in figure 1.15. It shows all the input and output ports utilized by the top module.

The simulated results of the decryption block are shown in figure 1.16. The cipher generated by the encryption block is decrypted to get the 8-bit message, which is found to be equal to the original message.

The on-chip power required to implement this decryption block was reported to be 1.191 W, as shown in figure 1.17.

A resource utilization summary for the implementation of the RSA decryption block on the FPGA board Virtex-5 XC5VLX2110T-2FF1738 is shown in

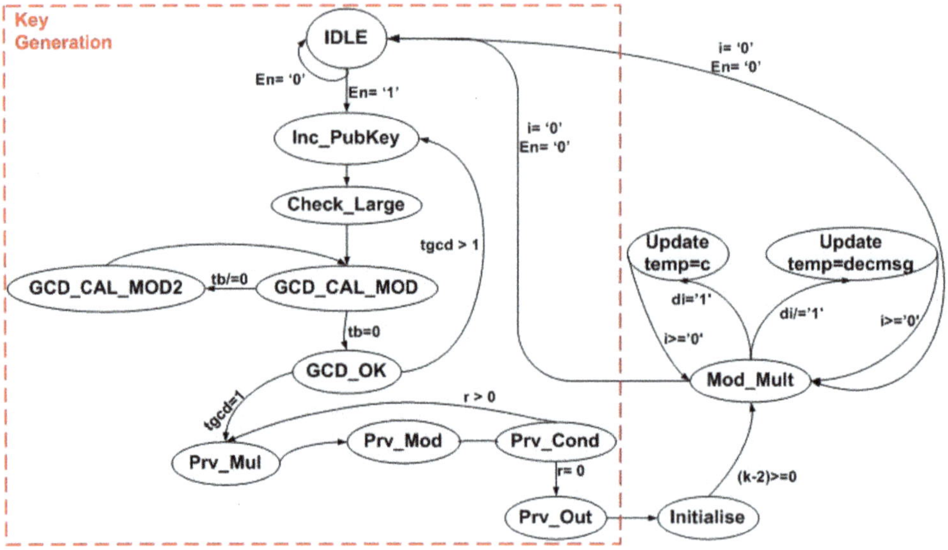

Figure 1.14. States involved in the implementation of the decryption block.

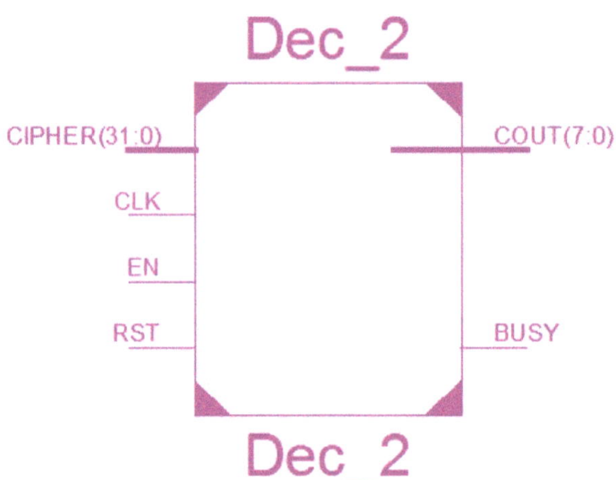

Figure 1.15. Top-level schematic of the decryption block.

figure 1.18. The total slice utilization of the decryption block was 6690, out of which a total of 371 slice registers were used.

1.5.4 Area and throughput comparison with existing implementations

The proposed design for RSA encryption was coded in VHDL and synthesized in ISE 14.7. Area and timing (frequency) information was generated and included in the synthesis report.

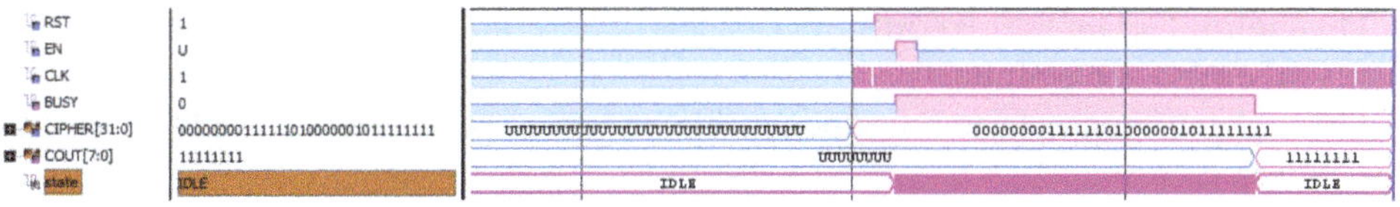

Figure 1.16. Simulated results for the decryption block.

Device	
Family	Virtex5
Part	xc5vlx110t
Package	ff1738
Temp Grade	Commercial
Process	Typical
Speed Grade	-2

Environment	
Ambient Temp (C)	50.0
Use custom TJA?	No
Custom TJA (C/W)	NA
Airflow (LFM)	250
Heat Sink	Medium Profile
Custom TSA (C/W)	NA
Board Selection	Medium (10"x10")
# of Board Layers	12 to 15
Custom TJB (C/W)	NA
Board Temperature (C)	NA

On-Chip	Power (W)	Used	Available	Utilization (%)
Clocks	0.009	1	---	---
Logic	0.000	6319	69120	9
Signals	0.000	8002	---	---
DSPs	0.000	7	64	11
IOs	0.000	12	680	2
Leakage	1.182			
Total	1.191			

Thermal Properties	Effective TJA (C/W)	Max Ambient (C)	Junction Temp (C)
	1.1	83.6	51.4

Supply Summary		Total	Dynamic	Quiescent
Source	Voltage	Current (A)	Current (A)	Current (A)
Vccint	1.000	0.856	0.009	0.847
Vccaux	2.500	0.130	0.000	0.130
Vcco25	2.500	0.004	0.000	0.004

	Total	Dynamic	Quiescent
Supply Power (W)	1.191	0.009	1.182

Figure 1.17. Decryption block power utilization summary.

Device Utilization Summary			
Slice Logic Utilization	**Used**	**Available**	**Utilization**
Number of Slice Registers	371	69,120	1%
Number used as Flip Flops	371		
Number of Slice LUTs	6,319	69,120	9%
Number used as logic	6,253	69,120	9%
Number using O6 output only	2,298		
Number using O5 output only	1,094		
Number using O5 and O6	2,861		

Figure 1.18. Device utilization summary for decryption.

1.5.4.1 Throughput calculation

The performance of a design also depends on the FPGA board it is running on. In the literature, different boards have been used by different authors. Directly comparing results based on slice utilization and frequency does not give a realistic idea of the comparative performances. Therefore, the efficiency parameter throughput/area is calculated separately for our implemented encryption and decryption designs.

For encryption

Maximum output required time after clock = T = 103.17 ns

Frequency = $1/T$ = 9.69 MHz

Total slice utilization = 1560 slices

Throughput = (size * frequency)/latency, where latency is the number of clock cycles taken by the output

Throughput/area = (size * frequency)/(latency * area) = (1024 * 9.69)/(1 * 1560)= 6.36 Mbps/slice

For decryption

Maximum output required time after clock = T = 92.171 ns

Frequency = 1/T = 10.149 MHz

Total slice utilization = 6690 slices

Throughput = (size * frequency)/latency, where latency is the number of clock cycles taken by the output

Throughput/area = (size * frequency)/(latency * area) = (1024 * 10.149)/(1 * 6690) = 1.66 Mbps/slice

Table 1.4 shows the a comparison between the results for the RSA encryption module and existing implementations. The table represents the on-chip area in terms of total slices utilized. Efficiency is compared in terms of throughput/area, where the throughput is calculated as the bit length multiplied by the frequency and divided by the latency.

Table 1.4 shows that the proposed designs for encryption and decryption have low resource utilization. Sufficiently secure designs utilize too much area, which results in a poor throughput-to-area ratio. The dashes in the table signify that the corresponding paper had limited information.

Our proposed design is both secure and has low area utilization. For encryption, the frequency obtained from our design is low, but the smaller areal utilization makes up for the high throughput/area ratio. The proposed design is cryptographically secure, power and area efficient, and is suitable for image applications.

1.6 Implementation results for the SPARTAN-6 (XC6SLX45-CSG324) FPGA board

To verify that our software simulations work on hardware, a hardware co-simulation was performed using the JTAG feature of the XSG. By running hardware co-simulation, a netlist was generated for the SPARTAN-6 (XC6SLX45-CSG324) FPGA board. Figure 1.19 shows the configuration of the SPARTAN-6 (XC6SLX45-CSG324) FPGA board with the computer. It shows the implementation results of the RSA 1024-bit algorithm on the FPGA.

We see three images on the computer screen. One of them is the input image, which is easy to understand. The encrypted image in the top right corner of the computer screen is generated by the system generator simulation. The image in the bottom right corner of the computer screen is generated by hardware simulation on the SPARTAN-6 (XC6SLX45-CSG324) FPGA board shown.

1.6.1 Design of the SPARTAN-6 (XC6SLX45-CSG324) FPGA board

Hardware co-simulation using a JTAG block was created and added to the Simulink model, as shown in figure 1.20. On executing this model, a hardware cosim file is

Table 1.4. Comparison with existing implementations.

Author	RSA size	Target device	Total slices utilized	Frequency (MHz)	Throughput/ area (Mbps/slice)	Power
Sharma [21]	1024	XC5VLX50–3FF676 (Virtex-5)	8012			
Shehata [23]	1024	XC5VTX240T-2FF175 (Virtex-5)	28 350	36.290	1.31	
Proposed (encryption)	1024	XC5VLX2110T-2FF1738 (Virtex-5)	1560	9.69	6.36	1.182
Proposed (decryption)	1024	XC5VLX2110T-2FF1738 (Virtex-5)	6690	10.149	1.66	1.19

Figure 1.19. Hardware implementation on an FPGA board.

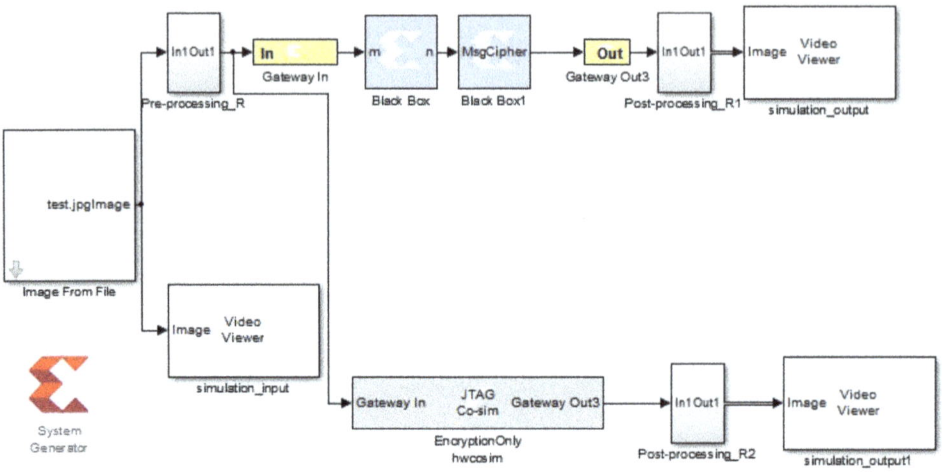

Figure 1.20. RSA 1024-bit algorithm implementation on a SPARTAN-6 FPGA board Simulink model.

generated. This file is the netlist generated for our board and is ready to be burned into the board.

1.6.2 Implementation of key security for the AES algorithm using RSA (hybrid cryptosystem)

Our proposed implementation can be used for applications such as the encryption of Quick Response (QR) code images. QR codes are used extensively in commercial applications, mainly advertising and financial transactions. Since they are public and QR readers are readily available on mobile devices, they can be scanned and read by anyone. An attacker can replace a QR code with a fake QR code and hence redirect the user to a phishing website that can steal the user's personal information. The status of a QR code, i.e. whether it is fake or not, cannot be determined until it is scanned. Therefore, to prevent any attackers from tampering with the QR code, secure encryption of the QR code is needed. Securing a QR code also ensures that it can only be read by the intended users and that not everyone gets access to the content. Symmetric algorithms, such as AES, are suitable if a QR code has to be scanned by a group of people (broadcast), but if it is secured by an asymmetric algorithm, the QR code will have to be encrypted exclusively for each receiver because each user has their exclusive private key. Also, symmetric algorithms are faster and less computationally intensive than asymmetric algorithms. However, existing symmetric-key algorithm implementations for image security lack the security, due to the use of the shared key. If the key is manipulated or leaked, it will result in the loss of important information. The shared secret key has to be secured strongly enough to prevent successful attacks on the key. Therefore, a model in which the QR code is encrypted and decrypted using the AES algorithm, and its shared key (a gray image) is secured by the RSA algorithm, is shown in figure 1.21 and its modeled system XSG is shown in figure 1.22.

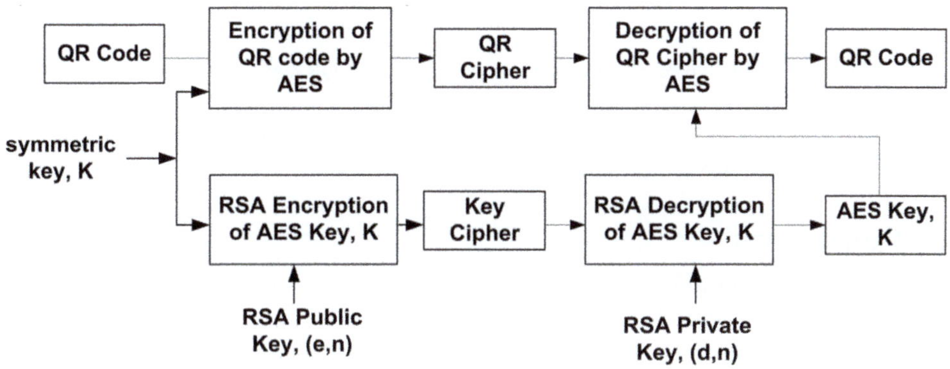

Figure 1.21. Security of a QR code and its shared key.

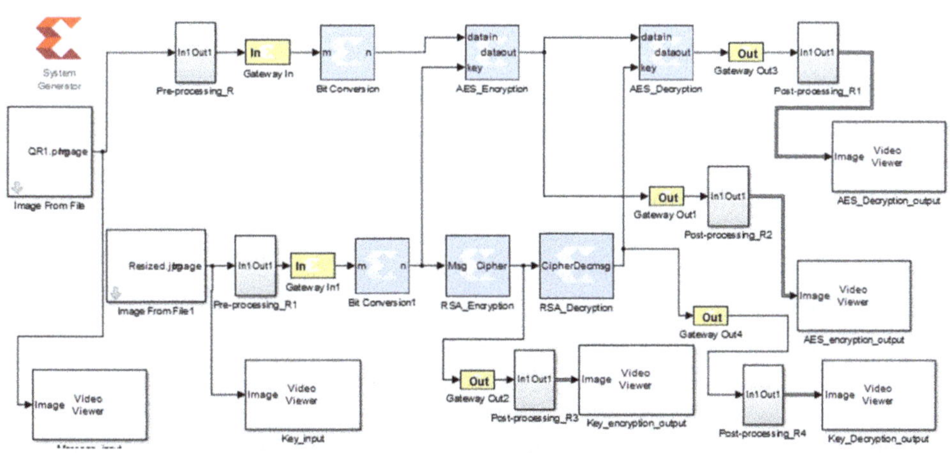

Figure 1.22. Model depicting encryption and decryption by an AES algorithm whose key is secured by the RSA algorithm.

(a) Shared key before encryption (b) QR code before encryption

(c) Encrypted key (d) Encrypted QR code

(e) Decrypted key (f) Decrypted QR code

Figure 1.23. Encrypted and decrypted image results for a QR code and its shared key.

Figure 1.23 shows the results of encrypting and decrypting a QR code and a key using the AES and RSA algorithms, respectively.

1.7 Conclusions

We proposed a state flow design for RSA-1024, for which the throughput-to-area ratio is improved while the size of RSA is kept at 1024 bits, so that the level of security is not compromised. This was achieved by a low utilization of the resources available on the FPGA board, thereby reducing the area significantly. The design was demonstrated for image applications using a system generator. A hybrid cryptosystem using an AES-RSA combination was also demonstrated on XSG for QR codes.

Acknowledgments

This work was carried out as personally sponsored research at the Electronics and Communication Engineering Department, the LNMIIT, Jaipur. The authors acknowledge the work done by MTech scholar Priyanka Gupta in carrying out the simulation work for this chapter.

Bibliography

[1] Sharma O, Saini S, Sharma A and Saini A 2016 A quadro coding technique to reduce self transitions in VLSI interconnects *IEEE International Symposium on Nanoelectronic and Information Systems (iNIS) (2016)* pp 98–101

[2] Gupta P, Saini S and Lata K 2017 Securing QR codes by RSA on FPGA *Int. Conf. Adv. Comput. Commun. Informatics* pp 2289–95

[3] Chhabra S, Jain H and Saini S 2016 FPGA based hardware implementation of automatic vehicle license plate detection system *Int. Conf. on Advances in Computing, Communications and Informatics (ICACCI)* pp 1181–7

[4] Rivest R L, Shamir A and Adleman L 1978 A method for obtaining digital signatures and public-key cryptosystems *Commun. ACM* **21** 120–6

[5] Leong P H W 2008 Recent trends in FPGA architectures and applications *4th IEEE International Symp. on Electronic Design, Test and Applications (delta 2008)* pp 137–41

[6] Chodowiec Pawełand Gaj K 2003 Very compact FPGA implementation of the AES algorithm *Int. Workshop on Cryptographic Hardware and Embedded Systems* (Berlin: Springer) pp 319–33

[7] Ruohonen K 2010 *Mathematical Cryptology* vol. 1 (Scotts Valley, CA: CreateSpace Independent Publishing Platform) pp 1–138

[8] Paar C and Pelzl J 2009 *Understanding Cryptography: A Textbook for Students and Practitioners* (Berlin: Springer)

[9] Koç Ç K 1995 *RSA hardware implementation* TR 801 RSA Laboratories

[10] Nedjah N and de Macedo Mourelle L 2002 Reconfigurable hardware implementation of Montgomery modular multiplication and parallel binary exponentiation *Euromicro Symposium on Digital System Design. Architectures, Methods and Tools (Dortmund, Germany, 4–6 September 2002)* pp 226–33

[11] McIntosh C 2007 Finding prime numbers: Miller Rabin and beyond *Furman Univ. Electron. J. Undergrad. Math.* **12** 1–4

[12] Mathur M and Kesarwani A 2013 Comparison between des, 3des, rc2, rc6, BlowFish and AES *Proc. of National Conf. on New Horizons in IT-NCNHIT* pp 143–8

[13] Surya E and Diviya C 2012 A survey on symmetric key encryption algorithms *Int. J. Comput. Sci. Commun. Networks* **2** 475–7

[14] Bonneau J and Mironov I 2006 Cache-collision timing attacks against AES *Cryptographic Hardware and Embedded Systems—CHES 2006 (Lecture Notes in Computer Science* vol 4249) (Berlin: Springer) pp 201–15

[15] Moghaddam F F, Alrashdan M T and Karimi O 2013 A hybrid encryption algorithm based on RSA small-e and efficient-RSA for cloud computing environments *J. Adv. Comput. Networks* **1** 238–41

[16] Abubakar A, Muhammad S J, Idrith Tijjani B, Zeki A, Chiroma H, Raji S and Mahmud M 2014 Cryptanalytic attacks on Rivest, Shamir, and Adleman (RSA) cryptosystem: issues and challenges *J. Theor. Appl. Inf. Technol.* **61** 1

[17] Menezes A J, Van Oorschot P C and Vanstone S A 2018 *Handbook of Applied Cryptography* (Boca Raton, FL: CRC Press)

[18] Lenstra A K and Verheul E R 2001 Selecting cryptographic key sizes *J. Cryptol.* **14** 255–93

[19] Karu P and Loikkanen J 2001 *Practical comparison of fast public-key cryptosystems* Telecommunications Software and Multimedia Lab at Helsinki Univ. of Technology pp 1–18

[20] Gaithuru J N, Bakhtiari M, Salleh M and Muteb A M 2015 A comprehensive literature review of asymmetric key cryptography algorithms for establishment of the existing gap *9th Malaysian Software Engineering Conf. (MySEC)* pp 236–44

[21] Sharma S and Zodpe H 2016 Implementation of cryptography algorithm for E-passport security *Int. Conf. on Inventive Computation Technologies (ICICT)* 3 pp 1–3

[22] Thobi A, Dhage S, Jadhav P and Chandrachood A 2015 Implementation of RSA encryption algorithm on FPGA *Am. J. Eng. Res.* **4** 144–51
[23] Shehata K, Hussien H and Yehia S 2014 FPGA implementation of RSA encryption algorithm for E-passport application *9th Int. Conf. on Electrical Engineering ICEENG* vol 9 pp 1–5
[24] Bo S, Kawakami K, Nakano K and Ito Y 2011 An RSA encryption hardware algorithm using a single DSP block and a single block RAM on the FPGA *Int. J. Netw. Comput.* **1** 277–89
[25] Nakano K, Kawakami K and Shigemoto K 2009 RSA encryption and decryption using the redundant number system on the FPGA *IEEE Int. Symp. on Parallel & Distributed Processing* pp 1–8
[26] Ghoreishi S S, Pourmina M A, Bozorgi H and Dousti M 2009 High speed RSA implementation based on modified Booth's technique and Montgomery's multiplication for FPGA platform *Int. Conf. on Advances in Circuits, Electronics and Micro-electronics* pp 86–93
[27] Sharma O, Saini A, Saini S and Sharma A 2016 a quadro coding technique to reduce self transitions in VLSI interconnects *IEEE Int. Symp. on Nanoelectronic and Information Systems (iNIS)* pp 98–101

IOP Publishing

Advances in Image and Data Processing using VLSI Design, Volume 1

Smart vision systems

Sandeep Saini, Kusum Lata, Abhishek Sharma and G R Sinha

Chapter 2

Modern ML methods for object detection

Abhishek Choubey, Shruti Bhargava Choubey, Anees Banu and Ipseeta Nanda

Today, image classification is a difficult task in the field of deep learning. Event recognition is a way of detecting physical reality in existing images or videos, such as those of living and non-living things. Real-time feature selection is now increasingly accepted in integrated transport tracking and monitoring equipment, new technologies, and Internet of Things (IoT) industries, as well as in self-driving cars, farming, security, and corporate implementations. The image classification device's controller has specific needs for legitimate, powerful solutions. Based on the application, platforms must have a large bandwidth, fast data transfer, and power efficiency.

Increasing power usage, algebraic sophistication, precision, and a high computational demand are indeed also difficulties for the electronic accelerators used in actual image classification. In the field of appropriate implementations of genuine image recognition, these challenges are the subjects of active and public current research. Multiple management strategies for the proper application of convolutional neural networks (CNNs) to image classification have also been presented in the past. Studies have attempted various types of data timing schemes, including quantification, to decrease computational cost and mathematical and memory complexity in dedicated hardware.

2.1 Introduction

Artificial Intelligence (AI) technology is a burgeoning branch of machine learning (ML) that involves computer programs and prediction using, for instance, the illustration of a shape from a supplied data set for retraining. ML development has been in progress for a long time; it uses a variety of models to examine data for diverse purposes. Recently, many ML applications have appeared, including Google Translate [1, 2], image analysis [3, 4], and voice control [5]. By 2021, more than ten

doi:10.1088/978-0-7503-3919-3ch2

terabytes of data will be created per second, thanks to the rapidly expanding use of digital cameras and the number of them in circulation [6]. ML has risen in popularity, both as a way to generate crucial results from a profusion of original information and also to express them using a sort of common model in the current age of big data. To keep up with the evolution in the processing of the acquired data, the computing capacity of all ML algorithms must be increased. The new AI approaches have given rise to personal computers that are clever in their perception and understanding of images of their environments, and now even better than humans at a variety of jobs. A computerized system, consisting of both techniques plus problem solving, offers us the capacity to train a robot to recognise the world through a sensor array. The field of computer vision seeks to enable computer systems to automatically see, identify, and understand the visual world, by simulating the way that human vision works [7]. Experimenters in computer graphics have hoped to create classifiers for tasks such as (i) machine vision, which determines whether a data set includes a particular item, (ii) object classification, which locates examples of semantic objects of a given class, and (iii) better picture understanding, which parses optimal threshold sectors for analysis.

Common object identification systems in machine vision may be split into two components: (i) the synthesis of features and (ii) classification. By establishing a loss function for the specified model, a training set may be used to develop weights and biases for classification.

The problems faced in the vision-based area are extremely tough, however, due to the wide variety of arithmetic used and/or the fundamental difficulty of recovering unknown factors from data that are insufficient to properly describe the method. Practically and theoretically, it is critical to investigate these issues. Classification in image recognition is a problem that uses cross-transfer learning to simplify the process of edge detection and identification (using CNNs). The goal of CNNs is to reduce a high-dimensional input picture to a low-dimensional yet highly abstracted semantic output. Strategies based on deep neural network (DNN) frameworks have introduced an enormous adoption of cloud-based vision into both data-based and electronics applications, thanks to the massive data from the Visual Geometry Group (VGG) architecture, modern general-purpose processors (CPUs), and graphics processing units (GPUs). CNNs have become the gold-standard computing engine in speech recognition in recent times. To ensure that CNNs can achieve near-human levels of accuracy in various visual recognition tasks, such as classification, tracking, and segmentation, a large number of denser and more dynamic interactions are created. However, this increased precision comes with significant computational costs. As a corollary, specialised players have emerged.

2.2 Machine-learning overview

ML is a science that seeks to 'learn' data and data knowledge using the processing power of 'the new machine,' or machines, as the title suggests. While the basic objective of deducing knowledge from data is comparable to that of learning mathematics, the integration of several procedures, considering the type of data on

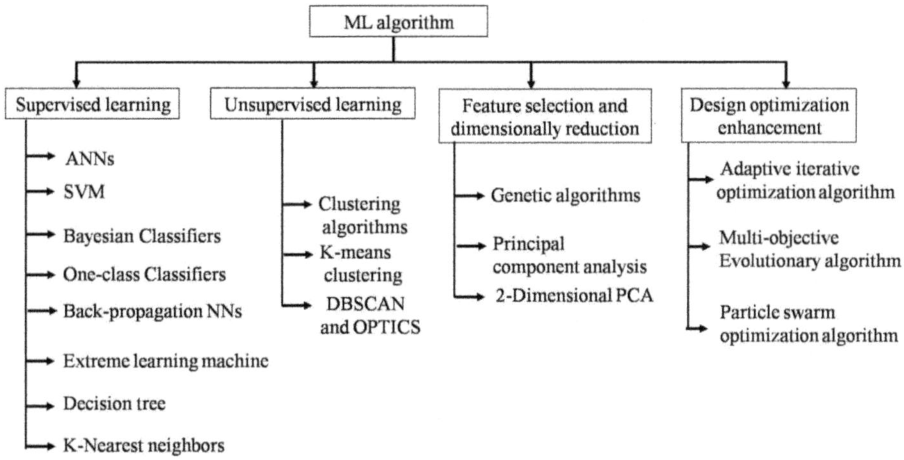

Figure 2.1. Existing ML methods.

which they are based, is more delicate. ML methods cover all the elements of learning mathematics, from geometry and computer science to a basic reproduction of pre-neural-network physiological mechanisms. The data sets involved can range from simple tabular, ordered statistics to more sophisticated structured data, such as movies and pictures, as well as unstructured information like graphs (figure 2.1). The following is the basic process for implementing ML algorithms:

(i) In the preprocessing phase, relevant features are first identified, and data that have these features are then retrieved from the original data so that they may be used to distinguish between the various desired output values. Following that, a variety of data cleansing and engineering features are available. Data augmentation is used to plan methods of attack, such as unique benefit preference, scaling, and creating a sample database for learning.

(ii) The learning algorithms needed are picked and run during the process of learning to retrieve the predictions from the training sample. These concepts are linked to a certain data concept. To reach the final forms, activities such as cross-validation, performance appraisal, and hyperactive modification are carried out.

(iii) The completed models are assessed in an evaluation step to see how well they perform against the test data. In practice, evaluation criteria may be selected and configured to fit a variety of circumstances.

(iv) In the modeling stage, the finished simulations are evaluated to determine the projected target correct output, given the current input data.

2.3 Computer vision algorithms

2.3.1 Image classification

The key field in which certain deep learning models play their most essential role in picture analysis is image categorization. In medicine, feature selection receives input pictures and generates an output categorization to determine whether or not an illness is present. Feature extraction uses a small piece of a medical picture as its input and provides two or more categorization categories. For improved categorization, local and global data are necessary.

2.3.2 Object detection

Feature extraction is frequently used in surveillance footage, intelligent automobiles, and the monitoring of different objects. In this chapter, we will cover the fundamentals of object tracking, as well as many of the deeper neural network models and several model techniques. Object detection is a computer vision approach that identifies things such as vehicles, buildings, and people, to name a few. Picture and video feeds are usually used to identify the items. Object detection is frequently used in video surveillance, intelligent automobiles, and the monitoring of body parts. In this chapter, we will cover the fundamentals of object detection, including some of the more widely used methods (figure 2.2).

Deep learning challenges aim to allow algorithms to effectively observe, analyze, and comprehend a field of vision in almost the same manner as humans [7].

Image analysis academics have aimed to create algorithms for the said sensory perception activities in the form of object recognition, which determines whether a picture input contains a certain item. This process involves: (i) scene interpretation, to parse a picture into units of meaning for analysis; (ii) object detection, to locate instances of semantic objects of a given class; and (iii) image classification, to locate examples of semantic objects of a given class.

Only a few years ago, classifying and locating an unknown number of distinct objects in a picture was thought to be an exceedingly tough task. Object detection is currently a viable task that has been used to identify anthropogenic greenhouse gas emissions by firms such as Google and IBM. Much of this development, however,

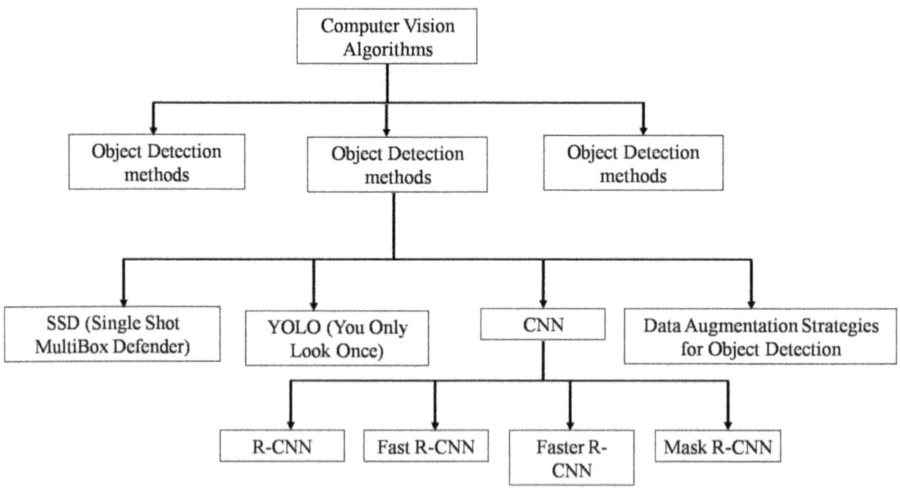

Figure 2.2. Computer vision algorithms.

was not simple! Object detection poses many more significant problems than those of picture categorization. We will take a second look at some of the intriguing challenges these difficulties provide, as well as some developing answers, after a quick introduction to the topic (see table 2.1).

The primary goal of object detection is to find significant things, create rectangular input images from them, and discern what class these items belong to. Detection is used in a variety of industries, including the recognition of humans by intelligent cars.

Over many years, this objective has required great academic effort, from Viola and Jones's facial identification algorithms in 2001 to RetinaNet, a speedy, high-precision, one-state detection network announced in 2017. The advent of CNNs was a watershed point in image classification history, since CNNs are now used in virtually every current implementation. However, the bulk of this chapter focuses on data mining techniques for image recognition, though other approaches face comparable problems.

2.3.3 Segmentation of images

Picture fragmentation, which is critical for computer vision, is defined as the division of an image into areas based on a set of criteria, with a requirement for the regions to be coherent and distinct. The ability to categorise images is a biologically essential human talent of the human visual system, a difficult but necessary function. The technique of segmenting one picture into numerous parts is known as image fragmentation. The term 'feature extraction' refers to the process of locating items and borders in pictures. For example, it can show the outcome of fragmenting a picture of a femur.

When you wish to validate the algorithm for a measurement method, you must first acquire data. Well over 100 pictures of the item for which you wish to create the model are required. Every photograph must also be of the highest standard and include one certain item (the one to be detected). Within the set of items, no item should be repeated; the greater the diversity of the photos, the broader the training. Shots containing only the objects are required for single-shot detection (SSD) and the faster region-based convolutional neural networks (R-CNNs), however in the case of You Only Look Once (YOLO), blank images are also utilised as classifiers. Here, we will demonstrate model training and the identification of a bespoke object, which is a company logo. In this method of measurement, the labeling of items in each video is mandatory. This method includes specific item information that has to be detected in the picture for object tracking, such as the x-axis, the y-axis, the width, and the breadth of the object inside the image. We utilise a labelling tool for the labelling, which enables users to create a pixel and collects the x-min, y-min, x-max, and y-max values and saves them in an XML file for each picture.

2.3.4 You Only Look Once (YOLO)

YOLO is a revolutionary technique for entity identification that predicts what things are while they are still in a picture using only one glance. Object identification is

Table 2.1. Challenges.

S. no.	Challenges	Description
1	Occlusion	Occlusion occurs when an object is partially or completely blocked by another object.
2	Scale change	When the size of an object varies a lot, object detection will fail in video tracking applications.
3	Motion blur	The moving object in a frame may be blurred and no longer looks the same.
4	Background clutter	When the background of an object has the same color or shape as the object, object detection may be erroneous.
5	Viewpoint variation	When the viewing angle of an object is changed, its shape and other features appear different. Thus, becomes difficult for object detection algorithms to identify that object.
6	Low resolution	Object detection stops working when an image has a low resolution and is pixelated.
7	Limited data	The limited amount of annotated data currently available for object detection proves to be another substantial hurdle. Object detection data sets typically contain ground truth examples for about a dozen to a hundred classes of objects. while image classification data sets can include upwards of 100 000 classes. Furthermore, crowdsourcing often produces image classification tags for free (for example, by parsing the text of user-provided photo captions). Gathering ground truth labels *along with* accurate bounding boxes for object detection, however, remains incredibly tedious work.
8	Class imbalance	Class imbalance proves to be an issue for most classification problems, and object detection is no exception. Consider a typical photograph; more likely than not, the photograph contains a few main objects and the remainder of the image is filled with background. Recall that selective search in R-CNN produces 2000 candidate region of interest polling (ROIs) per image—just imagine how many of these regions do not contain objects and are considered negatives.
9	Speed for real-time detection	Object detection algorithms need to not only accurately classify and localize important objects, they also need to be incredibly fast at prediction to meet the real-time demands of video processing. Several key enhancements over the years have boosted the speed of these algorithms, improving test times from the 0.02 frames per second (fps) of R-CNN to the impressive 155 fps of Fast YOLO.
10	Object localization and classification	The first major complication of object detection is its added goal: not only do we want to classify image objects but also to determine the objects' positions, generally referred to as the *object localization* task. To address this issue, researchers most often use a multitask loss function to penalize both misclassifications and localization errors.

framed as a regression issue in order to geographically separate region proposals and the linked classifier, which are forecast using a single data set from full pictures in one assessment; this is the first approach that completely bypasses the pipeline.

Convolution, normalization, pooling, and fully linked layers are the most common layers in CNNs as we know them:

(1) The convolutional layer can be used to identify higher-level characteristics from a source image.
(2) Max-pooling layer: this top-speed layer is a type of neural network.
(3) Batch normalization has been shown to be helpful in CNN training. It helps learning to converge faster and prevents overfitting of the network. The result of each convolution is adjusted by batch normalisation to decrease the internal alterations.

You Only Look Once, version 3 (YOLOv3) is a valid object recognition approach for recognizing specific things in movies, livestreams, and photographs. ML (AI) techniques use empirical systems to classify specific things in a group as important items. Unless otherwise specified, the systems classify objects in photos into groupings with comparable attributes and disregard the rest.

YOLO is an attempt to reduce the optimization technique designed for optimization problems; according to some of its competitors, it is not a traditional extractor that performs image classification as a byproduct. The input image is divided into M cells, which are responsible for five feature vector estimates that define the area around the object. It also provides a class label, which is a metric for the certainty that an object has been confined. As a consequence, the score has nothing to do with the sort of object in the box and everything to do with the box's shape. As with a traditional classifier, a classification is projected for each expected input image, yielding a normal distribution over all potential classes. The feature vector includes everything needed and the classification ratings are added together to provide an unique total score that reflects the chance that each container includes a specific access list. Only the panels with final scores larger than a certain threshold are retained, as the majority of the boxes have lower satisfaction values as a byproduct of these technical specifications. The YOLOv3 algorithm begins by dividing an image into grids. Each grid cell predicts the positioning of a specific number of input vectors (also known as anchor boxes) around things that perform well in the pre-determined categories.

Each bounding box has a confidence score that indicates how accurate the forecast should be, and each bounding box only identifies one item. The ground truth box dimensions are clustered to discover the most frequent shapes and sizes, and the boundary boxes are created by finding the most common shapes and sizes from the original data set. Because the training period of RetinaNet is longer than that of YOLOv3, the YOLOv3 average precision (AP) dictates a trade-off between speed and accuracy when utilising YOLO versus RetinaNet. However, with the use of a bigger data set, the accuracy of object identification in YOLOv3 can be made equivalent to that of RetinaNet, making it an attractive alternative for models that can be trained with huge data sets.

Common detection models, such as traffic detection, are examples of this, since the number of pictures of different cars is numerous, which allows for a large amount of data to be utilised to train the model. On the other hand, YOLOv3 might not be the best choice for specialised models, for which huge data sets are difficult to come by.

Here are some of YOLO's main benefits compared to other object detection methods:

- Extremely quick (45 frames per second, faster than real time)
- Tiny-YOLO is a smaller architectural version of YOLO that can function at a higher framerate (155 frames per second) with less precision than the original model
- Because the network recognises a generic object representation, real-world picture prediction and artwork are quite accurate
- YOLO object detection is open source and free.

2.3.5 Faster region-based convolutional neural networks

Deep learning approaches with a faster convolution and fast RCNN. Ren *et al* introduced the Faster R-CNN analyzer in 2015, just after the announcement of Fast R-CNN [8, 9]. Faster R-CNN (common object in context (COCO) mAP@.5 = 42.7%, COCO mAP@[.5,.95] = 21.9%, visual object detection (VOC)07 mAP = 73.2%, VOC12 mAP = 70.4%, 17 fps with ZFNet [4]). The invention of the region proposal network (RPN) was a significant new component of Faster R-CNN; it allows for nearly perfect feature maps. The majority of the independent parts of an activity recognition platform, including such proposal discovery, image segmentation, bounding box extrapolation, etc. have now been combined into a single, comprehensive system. Even though Faster R-CNN solves the efficiency restriction of Fast R-CNN, there is still some calculation repetition in the next detection phase. Several improvements have been suggested subsequently, including region-based fully convolution network (R-FCN) [3] and lightweight headed R-CNN [10].

The faster Area CNNS is another cutting-edge CNN-based computer vision object identification approach. This system transforms the input image into deep layers that produce a candidate solution. Instead of using a chosen selection method to discover the regional suggestions given in previous rounds, a separate network is used to learn and estimate such locations. The image is then classified inside the suggested region using a region of interest (ROI) mechanism, which forecasts the feature vector for the center pixel, altering the anticipated sector recommendations.

The region proposal network (RPN) instructional process uses a boolean identifier for every element; 1 identifies the presence of such a product and 0 indicates its disappearance; using this approach, any intersection over union (IoU) greater than or equal to 0:7 confirms the material's involvement and an IoU less than or equal to 0:3 indicates its disappearance.

2.3.6 Support vector machine (SVM)

The support vector machine (SVM) is a well-known machine-learning algorithm used for a range of text categorizations. SVM is well-known for its high prediction performance in domains such as facial recognition software, image processing, identification of characteristics, biochemistry, and tumor categorization. An SVM classifier is constructed using a neural network of data sets. The SVM classifier is then used to characterise any additional data using predefined network parameters created during the training period. SVMs have been proven to have higher accuracy than other approaches and feature in a number of current software solutions. They are powerful data mining techniques that excel at a wide range of supervised learning. This chapter highlights the SVM classifier, which successfully separates two classes of feature vector.

There are two major steps in a supervised approach such as an SVM: learning/ training and categorisation. During the training cycle, a taught system is constructed from an input test set, and a feature selection is established by distinguishing sequence data from two categories and using an optimal distance matrix. Support vectors (SVs) are random sets that lie on the decision function that are established in the previous training cycle and used in the analysis phase. Pipelining divides a process into smaller phases, allowing several activities to be completed at the same time. Pipelining improves data flow and speeds up the entire process. Many implementations employ the pipeline method, seeking to build efficient parallel pipelined designs by utilising the parallel processing capabilities of FPGAs. Three alternative configurations for implementing radial basis function kernel (RBF), polynomial, and sigmoid kernel functions are presented, as well as a completely pipelined architecture for implementing and speeding up SVM classification.

The FPGA's integrated digital signal processing (DSP) slices and block random access memories (RAMs) are used to create a new processor with a fully pipelined architecture. An SVM classifier with 760 SVs is implemented using 768 DSPs and 800 block RAMs using the suggested architecture. At 370.096 MHz, an elevated throughput of 2.89 106 periods per second is reached while categorising 128 feature dimensions.

2.3.7 Systolic array architectures

The systolic array structural design is a modular platform that may be configured to increase computation speed by combining parallelism and pipelining techniques. The systolic architecture is made up of a series of basic processors, or processing elements (PEs), that handle data flow and memory efficiently. For a variety of applications, in particular matrix-multiplication-based applications, a parallel FPGA is commonly utilised to create a systolic array architecture. The literature describes several SVM implementations that take advantage of the FPGA-based systolic array architecture to achieve faster parallel processing. By eliminating the use of computationally expensive multipliers, the multiplier-less method seeks to minimise hardware complexity. Many designs predicated on the multiplicative effect method have been established and documented in research papers. The [11] equipment feature vector is used in several of these designs. Compared to the existing Gaussian percentage increase kernel approach for boosted SVM [12], which is again based on a simultaneous synchronous systolic array architecture and predicated on a simplified multiplicative effect design, it offers a sufficient recognition rate (category B). To decrease hardware overhead and energy usage, the 'message conveyed' and 'add' operations utilise a simplified multiplier-free kernel. Furthermore, the canonic signed digit (CSD) and canonic sub-elimination (CSE) symbols are used in representation techniques that reduce the number of adders needed, reducing hardware complexity [13]. Three different developed filters (binary linearity, binary asymmetric, and word vectors) that employ the proposed CSD-oriented multiplicator kernel were compared with the standard vector multiplication kernel [12] on the basis of capacity utilization. Interestingly, the three developed classifiers saved 1, 2.7, and 3.5% of energy, compared to a traditional vector product classifier, a single-shot Mobilenet detector (SSD), as reported by Liu *et al* during 2015. It was a second one-stage detector. The creation of cross- and intra-intrusion detection is SSD's biggest contribution. It greatly improves the overall reliability of a single-stage detector, especially for small items. In terms of high detection accuracy, SSD has the upper hand (VOC07 mAP = 76.8%, VOC12 mAP = 74.9%, COCO mAP@.5 = 46.5%, and mAP@[.5,.95] = 26.8%), and an improved version works at 59 fps. SSD and previous detectors differ in that the former recognises objects of five different lengths at various levels of the web, but it only really discovers entities in the upper part.

2.3.8 Mask R-CNN

Mask R-CNN is a straightforward generalization of Rapid R-CNN in theory, but careful construction of the filter section is crucial for excellent results. Equally significantly, Quicker R-CNN was not even adjusted to resemble neural variables pixel by pixel. The way in which RoIPool, the de facto fundamental function for responding to examples, uses spatial resolution normalization for feature-based detection exemplifies this. The principle of Mask R-CNN is basic: for each contender item, the improved R-CNN provides two inputs: training data and a corresponding image offset; we add a third branch that produces the object mask. As a result, Mask R-CNN is a straightforward and obvious concept. The supplementary mask production, on the other hand, is unique with respect to the class and box results, necessitating the retrieval of an even finer spectral signature of an item.

Mask R-CNN uses the same two techniques and is equivalent in the first step (which is RPN). Mask R-CNN produces a binary image for each RoI in stage two, in addition to evaluating the category and the box offset. This is different from most current systems, which rely on masking estimates for categorization. The principle of Fast R-CNN, which performs linear and nonlinear zigzagging in parallel (which prove to greatly simplify the intra pipeline of classic R-CNN), is continued in this method.

A typical procedure for obtaining tiny region proposals from every RoI is RoIPool. RoIPool first formulates a floating-number RoI to the graph's characteristic periodic precision, then subdivides it into spatial bins, which are subsequently quantified, and then aggregates the evaluation metrics covered by each bin (usually by max pooling). Quantization is performed on continuous coordinate, for example, on a continuously variable *a* by setting [*a* = 16], where 16 is a feature vector phase and is averaging; quantification is also done when breaking it into bins. The RoI and the spectral information are misaligned as a result of these quantizations.

2.4 Hardware implementation

Deep learning techniques and complex problems are driving recent advances in image processing algorithms, but they also rely on location awareness in accelerated graphics, which provides the effective computational frameworks to enable the accurate instruction and execution of massive, dense, and multi-layered machine learning (figure 2.3).

Hardware acceleration uses personal computers to execute computations with less delay and greater capacity than traditional software implementations using general-purpose CPUs. Traditionally, von-Neumann-style architectures for solving quadratic equations (such as CPUs) have been designed largely for effective serial computation with complex task scheduling. When faced with deep CNN systems that demand parallel dense computing, high data maintainability, and wide processing power, they suffer from high energy consumption and poor physical memory for data transfer.

In practice, an operation can be calculated quicker by a hardware platform that has been built and/or specifically programmed for the purpose. By providing for high throughput and unique support for spatial changes as well as lowering the overhead of instruction management, hardware acceleration has enabled substantial customisation of processing capabilities.

(i) **Graphics processing units (GPUs)**

GPUs were initially designed to speed up image computing. GPUs are specifically designed to process bright lights, triangle configurations, and decoding. A GPU computer, also available in the form of cloud computers based on GPUs, is not only a strong visual accelerator but also a directly

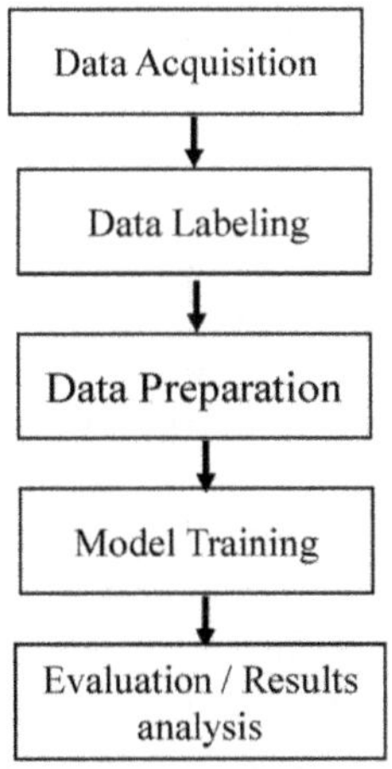

Figure 2.3. Standard implementation steps.

binding computational and storage processor, with mass spectrometry and the storage capacity for large parallel computations (general-purpose graphics processing unit (GPGPU)). Unlike multicore CPUs, which are generally out-of-order, multiprocessor CPUs that run at high speeds and use large caches to reduce separate latencies, GPGPUs are characterized by a large number of cores that run at lower speeds and rely on smaller caches. A wide range of coding languages, such as computational unified programming interfaces (Compute Unified Device Architecture (CUDA)) and Open Computing Language (OpenCL), have been investigated and used for GPU-accelerated embedded software, personal computer workstations, small business datacenters, virtualized social media, and slightly elevated computer technology (high-performance computing (HPC)) servers to deliver remote-graphics-card mobile apps using distributed systems.

For example, the Nvidia GeForce GTX and AMD Radeon HD GPUs are built for intensive gaming, whereas the Nvidia Quadro and Titan X series GPUs are meant for business workstations. Computational intelligence technology has just emerged, ushering in substantial advancements in GPU computing.

CNN, for example, can benefit from a parallel computational structure in the following ways:

(i) the regularization of an nn matrix with a k k kernel can be done in parallel;

(ii) subsampling/pooling can be implemented in different ways by running different accumulation operations on the data;

(iii) the activation of each neuron in a fully connected layer can be parallelized using a binary-tree multiplier. GPGPUs have been identified as a suitable fit for accelerating deep learning, due to their excellent parallel processing architectures and powerful floating-point capabilities. A variety of GPU-based CNN libraries, such as cuDNN, Cuda-convnet, and others founded on the popular deep learning framework, have been created to allow highly efficient CNN deployment on GPUs, for example, Caffe, Torch, TensorFlow, etc.

(ii) **Field-programmable gate arrays (FPGAs)**

While GPUs have been shown to provide exceptionally high capacity and are widely utilised to accelerate graphics in DNNs, their significant power output makes them unsuitable for energy/power-constrained applications, such as IoT devices. As a result, DNN acceleration is moving toward a more energy-efficient FPGA-based approach. FPGAs enable us to achieve variable scalability, customize data types, and implement hardware architectures, allowing us to adapt current DNN models using improved sparsity and tight network topologies with flexibility in mind. Furthermore, after manufacture, FPGAs may be reprogrammed for specific tasks and purposes.

The FPGA is a technology that consists of a structure of customizable interconnects that links a set of integrated circuit blocks. Analog-to-digital units (DSPs) for multiply–add–accumulation (MAC) operations, reference tables (look-up tables (LUTs)) for multimodal memory elements, and block RAMs allowing for database management are often found in today's FPGAs. Traditionally, an FPGA system has been defined using an object-oriented paradigm (hardware description language (HDL)) such as Verilog or the VHSIC hardware description language (VHDL) at the register-transfer level (RTL). To accurately define the detailed hardware implementation, this low-level design technique necessitates a significant amount of work and hardware knowledge. The incorporation of high concurrency between several hardware component high-level synthesis (HLS) tools has recently been completed to help FPGA design efficiency by leveraging high-level programming languages such as C and C++, and automatically compiling high-level descriptions to yield low-level specifications (i.e., HDL).

In reality, a DNN model is typically trained or fine-tuned on an elevated computing platform such as a GPU, and the FPGA speed is used for DNN inference to analyse the provided input data using the pre-DNN model.

The goal of the chapter is to develop an efficient translation from the pretrained sophisticated DNN model to the restricted hardware resources (high-density logic and memory blocks) available in FPGAs. The following hardware-friendly algorithmic optimizations were used to address this technological challenge: (i) algorithmic operation, (ii) data-path optimization, and (iii) model compression.

To minimise the number of mathematical functions used during inference, computational transformations such as general matrix multiply (GEMM), fast Fourier transform (FFT), and the Winograd transform can be used with feature maps and/or multilayer kernels. GEMM is a widely used method for data processing.

DNNs are neural networks that are used in CPUs and GPUs to vectorize the computations of parabolic and fully connected layers. FFT reduces the arithmetic complexity of 2D convolution by changing it into element-wise matrix multiplication. The data route is optimised by unrolling the convolution layers of CNNs and transferring them onto a limited number of PEs in order to fully leverage the concurrency.

DNNs are primarily utilised for error-tolerant applications and often include a large number of redundant parameters. As a result, a great deal of effort has been made to simplify DNN models, and, as a result, to reduce the complexity of their embedded systems. One simple method is to encapsulate each numerical value with the required word length based on its range. Alternatively, to give various scaling factors to distinct numerical parameters inside the same network, a dynamic method can be used. A DNN may employ binary weights and registrations in the extreme case, resulting in an exceedingly accurate recognition known as a binary neural network (BNN) [15, 33, 36, 37]. Because arbitrary addition and multiplication can both be accomplished using basic logic gates, a BNN may be assessed at a very cheap computational cost. A ternary DNN [38], unlike a BNN, sets its weights to 1, 0, or −1, enabling each weight to be encoded by two bits. Algebraic calculations in all neurons, on the other hand, are accomplished using floating-point arithmetic (FP32).

(iii) **Application-specific hardware accelerators**

To satisfy a given processing need, a typical computer system is frequently heterogeneous, consisting of a heterogeneous set of processors, such as CPUs combined with different processors. The influence of implementation details on knowledge processing can be seen in processors that work in tandem with CPUs. There are numerous hardware-based units in the form of either whole devices or coprocessors that are specifically created for computer vision and/or other AI workloads, in addition to the well-known coprocessors such as GPUs and FPGAs. The Google Tensor Processing Unit (TPU) is a whole device created specifically for neural networks and optimised for the Google TensorFlow framework. The TPU is designed to handle a large amount of low-precision (e.g. 8-bit) mathematics. Many Google applications, such as their search engine and AlphaGo, are already powered by it. The Intel Nervana artificial neural processor (NNP) was intended to even provide machine-learning primitives the versatility they need, while keeping the basic hardware components as efficient as possible.

Mobileye EyeQ is a line of systems on a chip designed for vision processing in self-driving vehicles. It has demonstrated the capacity to handle sophisticated and computationally expensive visual tasks while consuming minimal power. Various AI coprocessors for mobile devices have recently been created, and an arms race appears to be on the horizon. A Hexagon 685 DSP core in the Qualcomm Snapdragon 845 mobile CPU provides advanced on-device AI processing in image, speech, and gaming applications. The PowerVR Series 2NX/3NX NNA is one of Vision Technologies' neural network accelerators (NNAs).

2.5 Conclusions

Computer vision appears to be a hard study topic, but it has attracted a certain amount of publicity as a scientific subject. Modern imaging systems have developed considerably as a result of the introduction of big data, superior supervised neural techniques, and the acceleration of advanced computing.

In this chapter, we provided a detailed overview of object identification techniques. We highlighted recent advances in both methodologies for a wide range of computer tasks, such as image processing, object recognition, and edge detection, as well as promising embedded systems used to implement DNNs successfully in real situations, such as GPUs, FPGAs, and other future models of equipment accelerator.

Bibliography and further reading

[1] Derrien S and Rajopadhye S 2001 Loop tiling for reconfigurable accelerators *Conf. on Field Programmable Logic and Applications*

[2] Liu B *et al* 2015 Sparse convolutional neural networks *Conf. on Computer Vision and Pattern Recognition*

[3] Zhang X *et al* 2017 High-performance video content recognition with long-term recurrent convolutional network for FPGA *Conf. on Field Programmable Logic and Applications*

[4] Adam P *et al* 2017 SPARCNet: a hardware accelerator for efficient deployment of sparse convolutional networks *J. Emerg. Technol. Comput. Syst* **13** 31

[5] Liu L *et al* 2018 Deep learning for generic object detection: a survey arXiv:1809.02165

[6] Mittal S 2018 A survey of FPGA-based accelerators for convolutional neural networks *Neural Comput. Appl.* **32** 1109–39

[7] Garcia-Garcia A *et al* 2017 A review on deep learning techniques applied to semantic segmentation arXiv:1704.06857

[8] Donahue J *et al* 2014 Long-term recurrent convolutional networks for visual recognition and description *Trans. Pattern Anal. Mach. Intell.* **39** 677–91

[9] Kobielus J 2018 Powering AI: the explosion of new AI hardware accelerators *InfoWord*

[10] Kusum L and Saini S 2020 Hardware software co-simulation of an AES-128 based data encryption in image processing systems for the internet of things environment *2020 IEEE Int. Symp. on Smart Electronic Systems (iSES) (Formerly iNiS)* pp 260–4

[11] Perronnin F, Sánchez J and Mensink T 2010 Improving the Fisher kernel for large-scale image classification *European Conf. on Computer Vision* pp 143–56

[12] Liang S *et al* 2018 FP-BNN: binarized neural network on FPGA *Neurocomputing* **275** 1072–86

[13] Ma Y *et al* 2017 Optimizing loop operation and dataflow in FPGA acceleration of deep convolutional neural networks *Int. Symp. on Field-Programmable Gate Arrays* pp 45–54

[14] Yang T-J, Chen Y-H and Sze V 2017 Designing energy-efficient convolutional neural networks using energy-aware pruning *Conf. on Computer Vision and Pattern Recognition*

[15] Krizhevsky A and Hinton G 2009 Learning multiple layers of features from tiny images *Technical Report* University of Toronto

[16] LeCun Y *et al* 1998 Gradient-based learning applied to document recognition *Proc. IEEE* pp 2278–324

[17] Lowe D 1999 Object recognition from local scale-invariant features *Int. Conf. on Computer Vision* vol 2

[18] Dalal N and Triggs B 2005 Histograms of oriented gradients for human detection *Conf. on Computer Vision and Pattern Recognition* vol 1

[19] Oliva A and Torralba A 2001 Modeling the shape of the scene: a holistic representation of the spatial envelope *Int. J. Comput. Vis.* pp 145–75

[20] Li F-F and Perona P 2005 A Bayesian hierarchical model for learning natural scene categories *Conf. on Computer Vision and Pattern Recognition*

[21] Cortes C and Vapnik V 1995 Support-vector networks *Mach. Learn.* **20** 273–97

[22] Yang J *et al* 2009 Linear spatial pyramid matching using sparse coding for image classification *Conf. on Computer Vision and Pattern Recognition*

[23] Simonyan K and Zisserman A 2014 Two-stream convolutional networks for action recognition in videos *27th Int. Conf. on Neural Information Processing Systems* (Cambridge, MA: MIT Press) pp 568–76
[24] Wu Z *et al* 2015 Modeling spatial-temporal clues in a hybrid deep learning framework for video classification *23rd ACM Int. Conf. on Multimedia* pp 461–70
[25] Sharma S, Ryan K and Salakhutdinov R 2015 Action recognition using visual attention arXiv:1511.04119
[26] Li Z *et al* 2018 Video LSTM convolves, attends and flows for action recognition *Comput. Vis. Image Understand.* **166** 41–50
[27] Felzenszwalb P F *et al* 2010 Object detection with discriminatively trained partbased models *Trans. Pattern Anal. Mach. Intell.* **32** 1627–45
[28] Viola P and Jones M J 2004 Robust real-time face detection *Int. J. Comput. Vis* **57** 137–54
[29] Ahonen T, Hadid A and Pietikainen M 2006 Face description with local binary patterns: application to face recognition *Trans. Pattern Anal. Mach. Intell.* **28** 2037–41
[30] Everingham M *et al* 2010 The pascal visual object classes (voc) challenge *Int. J. Comput. Vis.* **88** 303–38
[31] Lin T-Y *et al* 2014 Microsoft COCO: common objects in context *European Conf. on Computer Vision* pp 740–55
[32] Girshick R *et al* 2014 Rich feature hierarchies for accurate object detection and semantic segmentation *IEEE Conf. on Computer Vision and Pattern Recognition* pp 580–87
[33] Uijlings J *et al* 2013 Selective search for object recognition *Int. J. Comput. Vis* **104** 154–71
[34] He K *et al* 2015 Spatial pyramid pooling in deep convolutional networks for visual recognition *Trans. Pattern Anal. Mach. Intell.* **37** 1904–16
[35] Girshick R 2015 Fast R-CNN *IEEE Int. Conf. on Computer Vision (ICCV)* pp 1440–8
[36] Achararit P, Hanif M A, Putra R V W, Sha_que M and Hara-Azumi Y 2020 Apnas: accuracy-and-performance-aware neural architecture search for neural hardware accelerators *IEEE Access* **8** 165319–34
[37] Jiang W, Yang L, Hsing-Mean Sha E, Zhuge Q, Gu S, Shi Y and Hu J 2019 Hardware/software co-exploration of neural architectures arXiv:1907.04650
[38] Hao C, Zhang X, Li Y, Huang S, Xiong J, Rupnow K, Hwu W-M and Chen D 2019 FPGA/DNN co-design: an efficient design methodology for IoT intelligence on the edge arXiv:1904.04421
[39] Zhang X *et al* 2020 Skynet: a hardware-efficient method for object detection and tracking on embedded systems arXiv:1909.09709
[40] Lu Q, Jiang W, Xu X, Shi Y and Hu J 2019 On neural architecture search for resource-constrained hardware platforms arXiv:1911.00105
[41] Jiang W, Yang L, Dasgupta S, Hu J and Shi Y 2020 Standing on the shoulders of giants: Hardware and neural architecture co-search with hot start arXiv:2007.09087
[42] Goel A and Bhujade R 2020 A prototype model for image encryption using zigzig blocks in inter-pixel displacement of RGB value *Int. J. Adv. Trends Comp. Sci. Eng.* **9** 6905–12
[43] Zhang X, Wang J, Zhu C, Lin Y, Xiong J, Hwu W and Chen D 2018 DNN builder: anautomated tool for building high-performance DNN hardware accelerators for FPGAs *2018 IEEE/ACM Int. Conf. on Computer-Aided Design (ICCAD)* p 56
[44] Hao C *et al* 2019 NAIS: neural architecture and implementation search and its applications in autonomous driving arXiv:1911.07446

[45] Chhabra S, Jain H and Saini S 2016 FPGA based hardware implementation of automatic vehicle license plate detection system *International Conference on Advances in Computing, Communications and Informatics (ICACCI) (Jaipur, India, 21–24 September 2016)*
[46] Marchisio A, Massa A, Mrazek V, Bussolino B, Martina M and Shafique M 2020 NASCaps: a framework for neural architecture search to optimize the accuracy and hardware efficiency of convolutional capsule networks *39th Int. Conf. on Computer-Aided Design* (New York: Association for Computing Machinery) p 114
[47] Feng X, Jiang Y, Yang X, Du M and Li X 2019 Computer vision algorithms and hardware implementations: a survey integration *VLSI Journal* **69** 309–20
[48] Wan S-Y, Lo L-J and Chang C-Y 2021 Disturbance-region removal on landmark-less registration of craniofacial images acquired by CBCTand 3D photography *Int. J. Emerg. Technol. Adv. Eng.* **11** 1–7
[49] Li Y *et al* 2018 A GPU-outperforming FPGA accelerator architecture for binary convolutional neural networks *J. Emerg. Technol. Comput. Syst.* **14** 18
[50] Geng T *et al* 2018 A framework for acceleration of CNN training on deeply pipelined FPGA clusters with work and weight load balancing *28th Int. Conf. on Field Programmable Logic and Applications (FPL)* pp 394–3944
[51] Sze V, Chen Y-H, Yang T-J and Emer J S 2020 Efficient processing of deep neural networks *Synthesis Lectures on Computer Architecture* (San Rafael, CA: Morgan & Claypool Publishers)
[52] Paliwal N, Vanjani P, Liu J-W, Saini S and Sharma A 2019 Image processing-based intelligent robotic system for assistance of agricultural crops *Int. J. Soc. Human. Comput. (IJHSC)* **3** 191–204

IOP Publishing

Advances in Image and Data Processing using VLSI Design, Volume 1

Smart vision systems

Sandeep Saini, Kusum Lata, Abhishek Sharma and G R Sinha

Chapter 3

Embedded intelligence for tracking facial expressions

Kyle Zampell and Ali Ahmadinia

In today's markets, facial detection technologies are making great strides forward in accuracy and speed, continuously pushing the limitations of computerized vision and offering ever-increasing opportunities for automation and machine learning, in what was once a human-dominated field. But while the accuracy of these systems increases, so do their computational requirements and complexity. Compounding the problem, advancements in hardware have not maintained the same rapid pace. Researchers now face the very real possibility that their algorithms will be bottlenecked by the very hardware it runs on. While we may be able to run current programs on current hardware, the idea of pairing a full-sized computer with a single camera does not scale well in terms of current or future applications. The goal of this chapter is to examine one component of facial recognition software: facial expression detection and evaluation, which aims to maintain a certain level of accuracy while running on hardware that is significantly weaker in computational power than a modern desktop workstation.

3.1 Introduction

The goal of this paper is to map a human face and mirror its expressions, using efficient hardware and algorithms to complete these tasks at a lower energy cost than their full-scale counterparts. In current technology, modern computers consider images to be two-dimensional arrays of three integer tuples. These arrays are purely data and values, and the tuples use a scale from 0 to 255 that represents the intensity of each of their associated colours (red, blue, and green). Zero translates to 'off,' the absence of light, and is drawn as black. The opposite represents the maximum brightness and can be drawn as white (or whichever colour has this value). This is important, as computers 'see' these data and do not consider that they could form

doi:10.1088/978-0-7503-3919-3ch3

shapes or objects. Humans, on the other hand, do not 'see' colours as a tuple of values; rather, we 'see' objects and shapes which happen to have these colours (which often differ from their surroundings). A simple illustration of this concept can be demonstrated by walking into an empty room. If the room has windows, a human would notice the rectangular shapes on what they perceive to be a vertical plane and could infer from the context of their location whether they were windows, doors, or posters. Our ingrained methodology of abstract thinking would have us perceive these rectangles to be windows before we bother to notice their colour or shape. It is this thinking, this processing of shapes instead of colours, which has been converted into applicable logic for fast image processing.

A large number of current research articles focus on the classifier-type processing of still images, which trains a virtual neural network to break down an image and tag it (and its occupants) with relevant information [2]. This type of evaluation is useful in many industries and applications, but it will not be the focus of this chapter. This research work was designed to use the concept that our subject(s) will need neither classification nor tagging, but rather locating and positioning. Furthermore, for this paper, we will be evaluating a video feed and not static images. For this type of requirement, OpenCV steps up to the plate, complete with a wealth of functionalities and features for image processing and evaluation.

OpenCV utilizes an established algorithm known as Haar Cascades, first described by Paul Viola and Michael Jones in 2001 [3]. The Haar Cascade is a sequence of feature identifications ranked by relevance and importance, used to quickly evaluate an image in order to locate regions where faces are most likely to be present. This sequence begins by isolating a smaller twenty-four by twenty-four pixel 'window,' which serves as the region of interest for the first pass of feature processing. This feature processing compares a binary mask against a subregion of four pixels, evaluating the brightness summation of all pixels in the 'on' region(s) against the summation of all pixels in the 'off' region. If this difference is negligible, then the feature is not present, and the cascade moves on to the next pixel location. If the difference is above a certain, trained threshold, then the classifier moves on to the next binary feature mask for evaluation. Using this sequence of roughly 6000 feature identifiers, the cascade can efficiently determine the location of a face with a 90% probability.

A similar open-source framework called OpenFace [4] takes a different approach, using a convolutional neural network (CNN) to identify faces based on a collection of trained features. OpenFace provides roughly the same functionality as OpenCV, in that, if given an image and tasked with searching, it will draw a box around each face it finds in the image. To accomplish this, a CNN is created using Torch (a publicly available GPU architecture codebase) [5]. Torch is designed to provide optimized support for advanced scientific calculations and machine learning algorithms built in Python. It is worth noting that OpenFace was designed with the original vision of matching the quality of Google's algorithm, FaceNet. As FaceNet is proprietary, we will not be exploring its inner workings or functionality, but it reportedly has the same conceptual design as OpenFace. It is noteworthy that at the time of writing this paper, OpenFace did not provide landmark data in its output for facial identification locations. While this would not normally be an issue, for this project's method of demonstrating the speed of the algorithm, it would be insufficient. Furthermore, OpenFace is designed to be powerful and comprehensive, but does not appear to have lightweight or low-computation applications in mind.

The last of the three main repositories which we will be covering in this paper builds on some of OpenCV's functionality and is called DLib [6]. DLib takes an alternative approach to facial recognition by using a combination of a histogram of oriented gradient (HOG), and support vector machine (SVM) techniques. HOG is a function which evaluates an image by dividing it into a grid, and then calculating the average slope of each grid square. The resulting array of slopes can be mapped into a collection of lines which enhance the dominant contours of each grid space. Visually displayed, this creates an arrangement of lines which can form visual contours that display where the computer believes the lines of the image are flowing. If one were to examine a circle evaluated in this manner using a grid size of two by two, the result would be a symmetrical diamond, in which each line would be tangent to the midpoint of the arc inside the individual box. By parsing this array of linear abstractions, the SVM seeks a specific collection/arrangement of lines which would most likely comprise the outline of the human face. Jung *et al* devised a motorized wheelchair control system that was tied to facial expressions captured and evaluated by a mobile phone [8]. This model was accurate, responsive, and simple in terms of implementation by the consumer, requiring only the installation of a small application on their smartphone and the connection of a small device to their wheelchair. We devised a real-time demonstration of speed, in the form of a telemetry-operated robot which was driven by specified facial expressions.

Building on the rising tide of crowdsourced computation, future implementations of real-time facial recognition are not likely to depend on a massive cloud cluster to process all incoming video and data feeds to locate persons of interest. Cameras are more likely to have an onboard processor which can handle a few known identities and which will be capable of processing its own images to find a match. If this results in a high enough probability of a match, the feed or data could be sent back to the cloud and tagged as such, reducing both the bandwidth used and the amount of raw processing capability required. This concept was inspired by the work of Sajjad *et al* [9] and their work on porting image processing algorithms to the Raspberry Pi platform.

The goal of this paper is to create a program which leverages modern computer image processing techniques to find a person's face in a video, track their orientation over time, and deliver commands to a robot which direct it to drive forward, speed up, turn left and right, and come to a complete stop. Most importantly, the proposed system will do this in real time, and be capable of running on a cheap microprocessor.

3.2 Description of algorithm

At the top level, the basic premise of this algorithm is to read in footage from a camera, process it to locate a single face in the image and then track the position and angle of this face for as long as it remains in the scene. To do this, the algorithm can be broken into several pseudo stages which mark points of progress:

Initialization
Image rotation

Facial recognition
Feature mapping
Feature evaluation
Feature rotation
Broadcast controls

Our initialization stage is the part of the algorithm that requires the longest execution time, but it only needs to run once, on initial startup. This step establishes a secure shell (ssh) connection to the robot, loads the Haar Cascades into memory, and configures the camera for video capture.

The image rotation step is exactly as the name suggests. The algorithm acquires the most current frame from the camera, creates a gray scale copy of the colour image, and then rotates it based on an internal rotation tracker value. When this algorithm is first initialized (and after a certain number of frames with no tracking), this rotation angle is set to zero degrees. This logic is based on the reasoning that the camera will always be oriented upright, and that the human subjects are unlikely to enter or constantly remain in frame with their heads tilted significantly in either direction. In a scenario in which the angle is a non-zero value, the image is rotated by the calculated number of degrees which re-orients the subject's face to make it upright.

The facial recognition step leverages the OpenCV Haar Cascades with a function call that returns a bounding box around any faces it detects in the image. In some situations, this function call can return two differently sized boxes for a single face. To handle this, our algorithm averages both boxes to produce a single set of boundaries and counts how many boxes were detected. This count must evaluate to a value greater than zero to allow the rest of the algorithm to continue processing the image. This check prevents several errors that would arise as well as allowing the algorithm to speed up its execution. If this count sequentially evaluates to zero for a specified number of times, the rotation angle of the image is set to zero, and the counter is reset.

The feature mapping stage originally negated the need for a facial recognition step; however, our testing found that this stage is very dependent on the size of the input to be evaluated. In order to counteract DLib's speed issue, our approach was to split the two stages in order to allow DLib to process fewer pixels. This step takes the box location found in the facial recognition step and then extracts that rectangle from the image as a subimage. This subimage is processed by DLib much faster than if it were handed the whole image, and returns an array of (x, y) coordinates that identify forty predefined features that outline a human face. These features include eyebrows, nose, eyelid openings, the inner and outer boundaries of the lips, and the full jawline. Once these features are identified, their coordinates are calibrated to match those of the full-sized image.

The feature evaluation stage is tasked with applying Euclidean and geometrical analyses of the feature coordinates in x–y space. In this step, the current rotation of the subject is evaluated based on the relative coordinates of the features provided by the feature mapping stage. This calculated rotation is then added to the algorithm's internally tracked rotation angle. The tracked rotation angle will continue to be updated for each image by the previous image, in an attempt to keep track of the subject in the frame. The algorithm then evaluates the space between two set points on the subject's lips, which returns a scalar value in pixels. The scalar lip value will be transferred to the broadcast controls step, while the current rotation angle will be utilized in the next stage.

The feature rotation stage was not initially planned but became a necessity due to the observed technical limitations of DLib and OpenCV. During testing, we found that if a subject's head was tilted by more than five degrees left or right, the detection rate dropped off dramatically. If the subject rotated by more than ten degrees, detection became impossible. To counter this, the feature evaluation stage determines a current rotation angle in degrees. This step then takes this angle and transposes the coordinates such that their relative position on the subject would remain the same when placed on the original, unmodified source image. This allows the feature mapping results to be displayed in a format which the subject can use to confirm their validity.

The last stage, broadcast controls, was designed to show the real-time speed at which this algorithm is executed. This step first evaluates the current angle and, if its absolute value is above a specific threshold, it sends a command to turn the robot. If the angle does not surpass this threshold, it then sends a speed command based on the scalar value mentioned in the feature evaluation stage. If that scalar is below a threshold, the algorithm sends a stop command. Due to the difference between the execution time per frame of this algorithm, and the relative time required for the robot to receive and process a command, this stage only executes once per a certain interval of processed frames.

Upon receiving a command to terminate, the algorithm sends a termination command to the robot, which ceases to run the local program, and closes the camera feed. This command is only checked for when the algorithm has completed the broadcast controls stage, and is the last action taken before the next frame is captured from the camera.

3.3 Implementation

3.3.1 OpenCV with three simple Haar Cascades

Initially, a prototype algorithm was designed which could read images from a camera in real time, find subject faces using the Haar Cascades, and then draw coloured boxes around them overlaid on the source image being displayed. This prototype was also able to measure the elapsed time taken to process each image, providing a benchmark for performance on a modern computer.

```
face_cascade = cv.CascadeClassifier('Lib/site-packages/cv2/data/haarcascade_frontalface_alt2.xml')
ret, frame = cap.read()
gray = cvtColor(frame, cv.COLOR_BGR2GRAY)
face = face_cascade.detectMultiScale(gray, scaleFactor = 1.5, minNeighbors = 5)
```

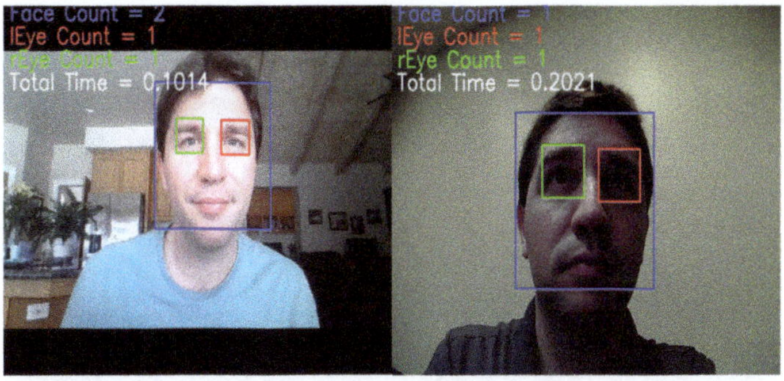

Figure 3.1. Comparison of OpenCV design with three Haar Cascades (left: laptop; right: Raspberry Pi).

Shown above is a sample of the process for reading in an image from the camera (cap), converting the colour image (frame) to a gray scale (gray), and then generating a bounding box (face) around each face found in the image by the classifier. The scaleFactor and minNeighbors values were the results of trial and error combining multiple values.

Once this was completed, a Raspberry Pi unit was set up to have the same libraries and configurations. Once the prototype was operational, we were able to perform a baseline comparison benchmark between the two platforms. In this baseline, we implemented three of the sample Haar Cascades: one for the frontal face, one for the left eye, and one for the right eye. On the first run through, we noted that the two eye cascades were tracking both eyes (that is, the left eye cascade tracked both the left and right eyes) and background objects, leading to the implementation of a filter. In this filter, we rejected any detections made outside the rectangle representing the face and any eye detections which were outside of their relative 'half.' The result looked like the image below, in which the left part is the post-processed frame from the laptop execution, and the right part is the post-processed frame from the Raspberry Pi execution (figure 3.1).

3.3.2 DLib

After we established this baseline, we began to experiment with methods of improving performance, accuracy, and memory footprint. However, OpenCV did not provide the level of detail we would need; all that it could provide was a coordinate and the lengths of two sides of a rectangle which would encapsulate a region it found. Another facial recognition package called DLib [6] provides forty coordinates arranged around key features of the human face. This library also proved difficult to implement, as the ready-made package was designed for C++, and their Python version had to be compiled on the target device.

As this was new ground, we experimented by running DLib and measuring how it impacted performance. Noticeably, frames began to take much longer to evaluate on the Raspberry Pi, which was shaping up to be an issue, considering that we wanted to add more functionality per frame. To achieve a better frame time, we tested different changes to determine why it was slower, and how we could address

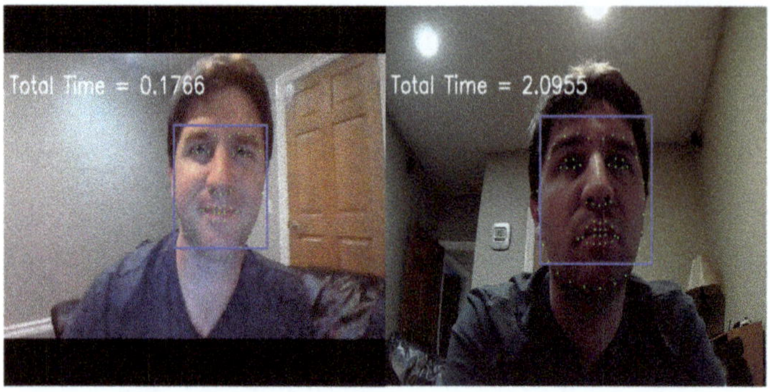

Figure 3.2. Comparison of DLib design results (left: laptop; right: Raspberry Pi).

that. These tests included altering the camera resolution, changing the ambient lighting in the scene, and using different subjects in the scene. We found that a direct light source aimed at the subject from behind the camera increased the accuracy, but not the frame rate. Altering the image resolution did have an impact, one that had a positive correlation with the size. When we lowered the camera's resolution to 640 by 480 (also called 480p), as we began to have issues at lower resolutions (with a few first occurring in 480p). Below is a comparison of the DLib functionality for a 480p image; the left part is the post-processed frame from the laptop and the right part is the post-processed frame from the Raspberry Pi (figure 3.2).

3.3.3 OpenCV and DLib

The reason for the slowdown was the number of pixels that DLib had to process. DLib took much longer than OpenCV to complete its calculations, but DLib was also the only one of the two that returned the data we required. Thus, we had the notion of tasking OpenCV with locating the face as a bounding box and then feeding that subimage to DLib for feature processing. This meant that the region found by OpenCV would be selected and fed to DLib as a separate image for processing. The results were a drastic improvement, as seen in the comparison below; the post-processed frames from the laptop are shown on the left and those from the Raspberry Pi are shown on the right (figure 3.3).

This resulted in a substantial increase in performance, cutting the time used per frame down to 0.13 seconds on the Raspberry Pi. It was also at this stage of development that we began to see some issues with OpenCV's box, namely, that it sometimes did not include the entire head of the subject. In these scenarios, DLib would not be successful, as some of the features would be outside of the provided image. The only remedy for this was manual intervention, in which we simply multiplied the dimensions and expanded the size of the bounding box. While this resulted in a near-negligible difference in the run time, it removed the risk of encountering the small box issue. Shown below is a sample of the lines required to implement and utilize DLib:

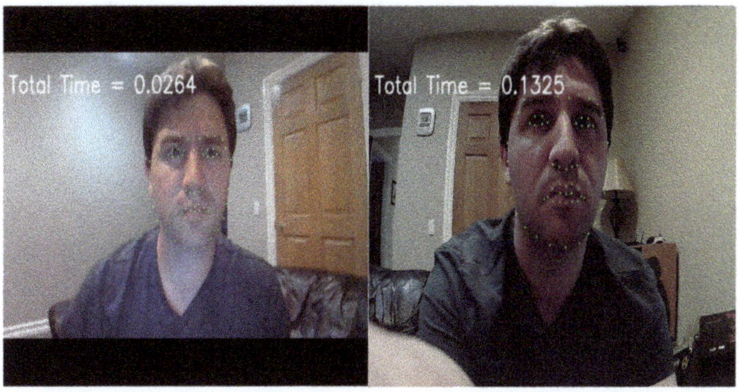

Figure 3.3. Comparison of the results of the hybrid OpenCV+DLib design.

```
predictor = dlib.shape_predictor('shape_predictor.dat')
rect = dlib.rectangle(startX, startY, endX, endY)
shape = predictor(gray, rect)
shap = face_utils.shape_to_np(shape)
```

In order of appearance, the variables are: startX, startY, endX, and endY (the boundaries provided by OpenCV), rect is a rectangular object structure, gray is the gray scale image taken from the camera, and shape is the array of *x–y* coordinate pairs for each facial feature. This shape array is where all the points of interest are stored.

3.3.4 Mitigating loss of track

We began to face the next issue after testing how well the algorithm found facial features with different expressions, poses, and lighting angles. There was a weakness in the identifiers of both libraries when the subject simply tilted his head too far to the left or right. OpenCV thankfully had multiple variants of image rotation functions built in, but we still faced an issue with verification. The image below best shows that even though the rotated image allowed the libraries to find the subject again, the coordinates were incorrect when applied to the original image. Note that the image below was taken on the laptop using the hybrid OpenCV + DLib design; the green dots represent the detected *x–y* coordinates found in the rotated image (figure 3.4).

Neither OpenCV nor DLib had a solution for this problem; therefore, we added functionality to enhance the time-series tracking capabilities of DLib. We devised a simple function that would accept two pairs of coordinates and a rotation angle, returning a new set of coordinates. This function first converted one coordinate (the target point) into polar coordinates, using the second coordinate as the origin point. The function then rotated this imaginary vector in polar space by the provided rotation angle and calculated the new *x–y* coordinate point. Below, we show how we calculate the tilt of the subject's head in the frame, where jaw is an array of 16 *x–y*

Figure 3.4. Demonstration of image rotation vertex mismatch.

coordinates, and these coordinates span from the top left of the jawline down around the chin and back up to the top right of the jawline:

```
def calcTilt(jaw):
L = dist.euclidean(jaw[0], jaw[8], [1.0, 0.0])
R = dist.euclidean(jaw[16], jaw[8], [1.0, 0.0])
vert = dist.euclidean(jaw[0], jaw[16], [0.0, 1.0])/2
hor = dist.euclidean(jaw[0], jaw[16], [1.0, 0.0])/2
hyp = math.sqrt(vert * vert+hor * hor)
tilt = int(math.degrees(math.acos((hor * hor + hyp * hyp -
vert * vert))/(2.0 *hor * hyp)))
IF L < R
  Tilt = -tile
Return tilt
```

With this tilt angle, the algorithm can properly rotate the input image so that the subject's face is near-vertical for optimal detection. When we compared the rotated image feature point (red) against the unrotated image with the newly calculated location for said feature point (green), the feature accuracies were virtually identical; the post-processed images from the laptop output are shown on the left and the rotated input gray scale is shown on the right (figure 3.5).

Our implementation ended up being context aware of where the point was located relative to the polar origin. First, the function translated the coordinates to a polar radius and angle of the horizontal axis and then assumed that the radius was in the positive-x positive-y quadrant for the remaining calculations. Then, after increasing/decreasing the angle, it determined whether this new angle was greater than ninety degrees or less than zero, in which case a special case would take effect. The function then repositioned the new polar vector to align with the original input's vector and converted that new vector into a pair of x–y coordinates.

```
def calPoint(x, y, theta, lastCenter):
R = math.sqrt((x-lastCenter[0])**2 + (y-lastCenter[1]**2))
beta = math.degrees(math.atan2(abs(y-lastCenter[1]),
abs(x-lastCenter[0])))
if x > lastCenter[0] and y < lastCenter[1]:
   alpha = beta + theta
```

Figure 3.5. Demonstration of rotated vertex alignment (left: laptop output; right: rotated gray scale).

```
    xMult = 1
    yMult = -1
  elsif x < lastCenter[0] and y > lastCenter[1]:
    alpha = beta + theta
    xMult = -1
    yMult = 1
  else:
    alpha = beta - theta
    if x > lastCenter[0]: #thus y > 1C
      xMult = 1
      yMult = 1
    else: #and thus y < 1C
      xMult = -1
      yMult = -1

  newX = lastCenter[0] + xMult * int(R*math.cos(math.radians
(alpha)))
  newY = lastCenter[0] + yMult* int(R*math.cos(math.radians
(alpha)))
  return (newX, newY)
```

Using these corrected coordinates, we created a two-step evaluation of the computed feature points, to output a turning rate and a drive speed for the live demonstration. For this, the first function takes the topmost two points of the jawline, along with the bottom centre point of the jawline. Using relative differences in the *x* and *y* coordinates of these points, it calculates a tilt angle (either clockwise or counterclockwise) and returns this angle (a negative value represents clockwise rotation of the image). The second function uses the centre pixels of both the upper and lower inner lip lines. It calculates the absolute pixel distance between the two and returns that value as a scalar. We show the final output image below, including the measurements of the features for the drive system (figure 3.6).

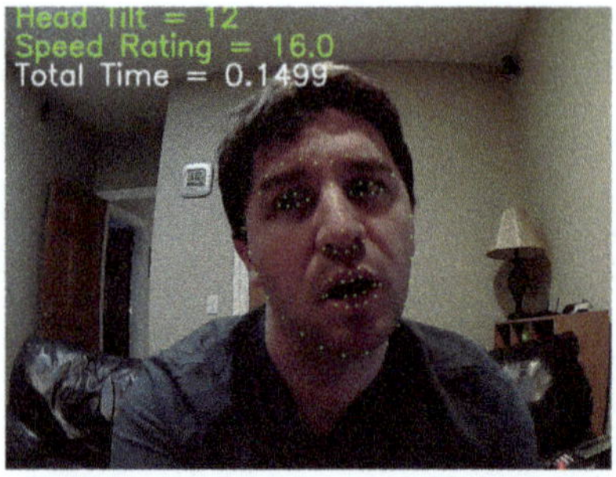

Figure 3.6. Raspberry Pi final output image with rotation and feature evaluation.

3.4 Results and evaluation

Overall, it was to be expected that the laptop would outperform the Raspberry Pi; however, the goal of this study was to experiment with how much capability could be compressed into such an economical package. While the laptop is currently selling for $1000, the Raspberry Pi unit is selling for $40. The laptop's specifications are as follows: an i7–7700HQ (quad core) at 2.80 GHz with 16 GB of ram. The Raspberry Pi's specifications are as follows: a quad core at 1.20 GHz with 1 GB of ram. As there was no accessible video footage that had been used in related work, we generated some which was seventy-five seconds long, featuring various facial expressions, different degrees of head tilting, and a left and right rotation to demonstrate loss of track when possible. This video was then fed through all four setups on each machine, tracking the frame number, the execution time for the frame, and whether or not a face was detected in the frame. We created the video as a control test, with a flat colour background and no other objects in a scene to prevent false detections.

The first comparison is that of the baseline OpenCV code. In this test, the laptop performed at an average rate of 0.05 seconds per frame, and the Raspberry Pi performed at an average rate of 0.19 seconds per frame. Given that the laptop has a clock cycle time of less than half that of the Raspberry Pi, this result is to be expected. Furthermore, the laptop had a detection rate of 67.90% and the Pi had a detection rate of 68.09%. These data are shown below in figures 3.7 and 3.8, respectively. Figure 3.6 represents the time per frame for total processing in seconds, and figure 3.7 shows whether the device performed a detection in a frame. Note that in figure 3.8, the loss of detection events appears to be synchronized.

The second comparison is that of the baseline DLib code. In this test, the laptop performed at an average rate of 0.28 seconds per frame, and the Raspberry Pi performed at an average rate of 2.05 seconds per frame. This was a drastic departure for the Raspberry Pi, compared to the OpenCV benchmark, taking almost nine times longer to complete. In return, the DLib implementation provided more usable information, such as the angle and orientation of the subject. It was also observed that the detection rates were 89.30% for the laptop and 89.76% for the Raspberry Pi. These data are shown below in figures 3.9 and 3.10, respectively. Figure 3.9 represents the time per frame for total processing in seconds, and figure 3.10 shows

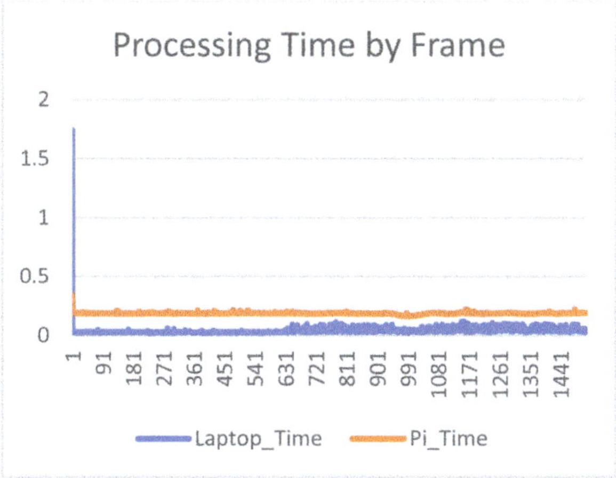

Figure 3.7. Processing time per frame for the OpenCV design.

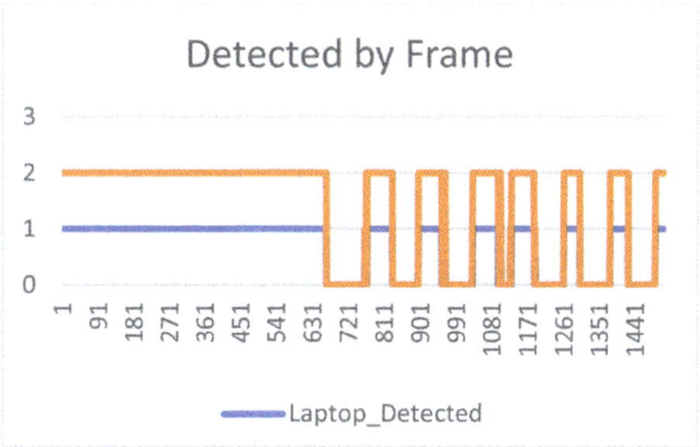

Figure 3.8. Stacked detection by frame for the OpenCV design.

whether the device performed a detection in a frame. Note that in figure 3.10, the loss of detection events again appears to be synchronized.

The third comparison is for the hybrid program, which utilizes both OpenCV and DLib in conjunction. In this test, the laptop performed at an average rate of 0.02 seconds per frame, and the Raspberry Pi performed at an average rate of 0.09 seconds per frame. With this implementation, the concept is far more practical. It should be noted that this hybrid setup has an even faster execution time than that of the baseline OpenCV design. It was also observed that the detection rates were 67.90% for the laptop and 68.10% for the Raspberry Pi. These data are shown below in figures 3.11 and 3.12, respectively. Figure 3.11 represents the time per frame for total processing in seconds, and figure 3.12 shows whether the device performed a detection in a frame. Note that the first frame(s) in figure 3.11 have a significantly higher processing time on the laptop, but only a slight uptick on the Pi. The code used is completely identical, and this anomaly occurred in every test performed on the laptop.

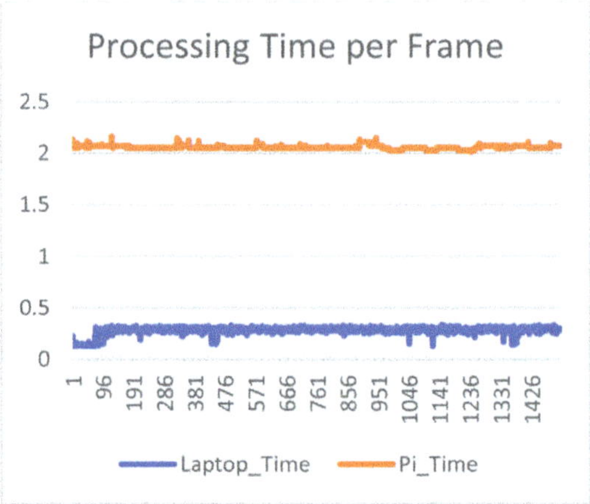

Figure 3.9. Processing time per frame for the DLib design.

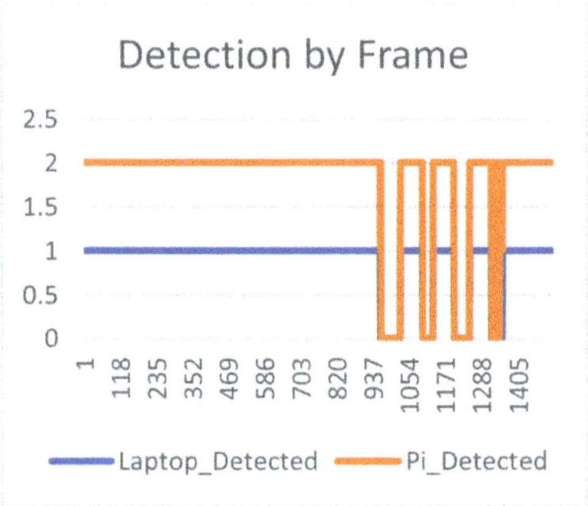

Figure 3.10. Stacked detection by frame for the DLib design.

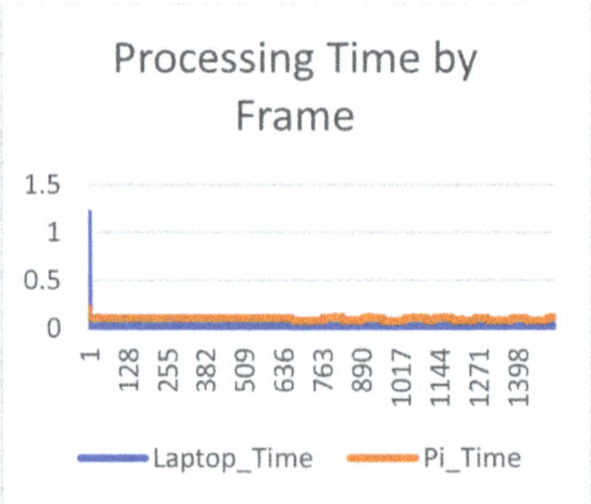

Figure 3.11. Processing time per frame for the hybrid design.

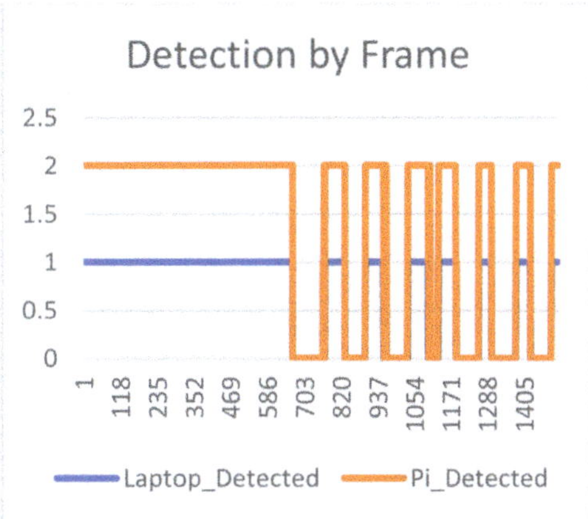

Figure 3.12. Stacked detection by frame for the hybrid design.

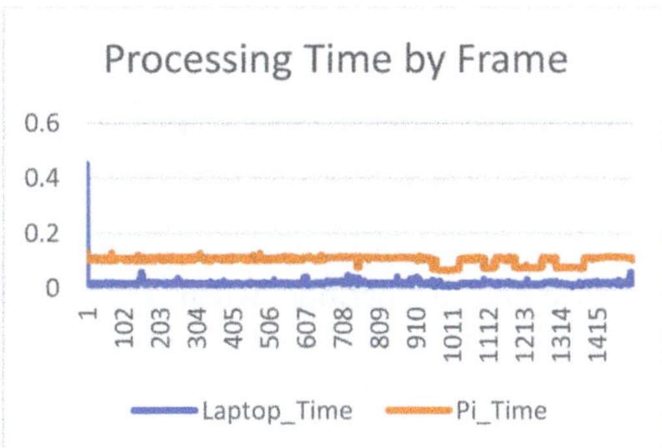

Figure 3.13. Processing time by frame for the hybrid design with translation.

The fourth comparison is for the hybrid program, including the point translation. In this test, the laptop performed at an average rate of 0.02 seconds per frame, and the Raspberry Pi performed at an average rate of 0.10 seconds per frame. My fears that this algorithm would be too computationally time-consuming were assuaged, as this small device is clearly capable of processing and evaluating an image feed at the rate of roughly ten frames per second. It was also observed that the detection rates were 82.43% for the laptop and 82.50% for the Pi. These data are shown below in figures 3.13 and 3.14, respectively. Figure 3.13 represents the time per frame for total processing in seconds, and figure 3.14 shows whether the device performed a detection in a frame. Note that there is a significant spike again in the processing time during the initial frame(s), as discussed for the third comparison.

Overall, from these four comparisons, we can argue that our design is competitive in terms of both time and accuracy. For comparison, the times per frame for the OpenCV, Hybrid_No_Tilt, and Hybrid (with translation) designs running on the Raspberry Pi are compiled in figure 3.15. Note that the Hybrid_No_Tilt design has

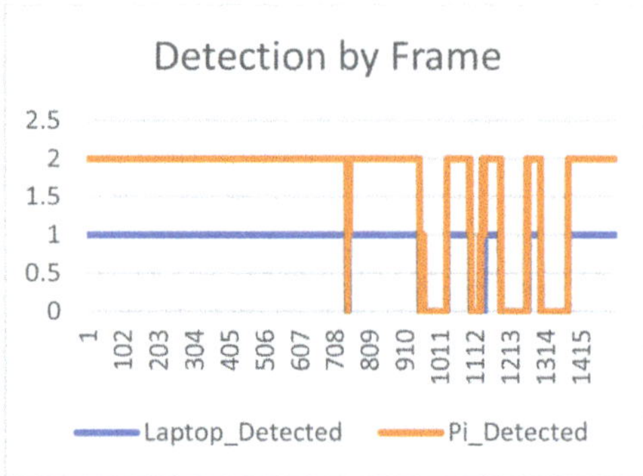

Figure 3.14. Stacked detection by frame for the hybrid design with translation.

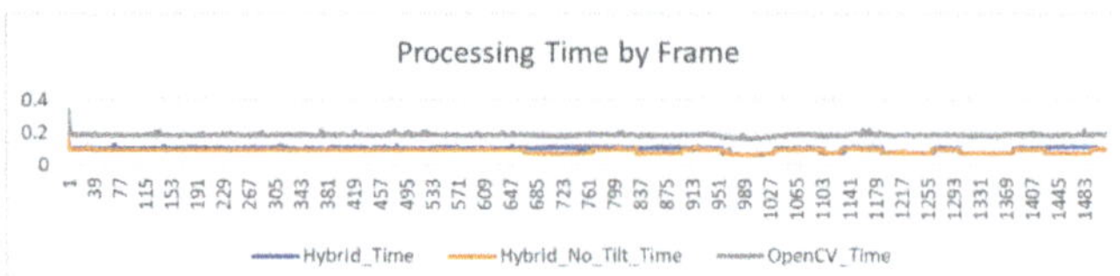

Figure 3.15. Processing time by frame for Raspberry Pi, comparing the OpenCV, hybrid, and hybrid with translation designs.

some drops in the time-per-frame value for periods in which the Hybrid design does not. These periods correspond to periods when the translation and rotation functions keep track of the subject during head tilting, unlike the other dips, in which both lose track due to head rotation.

3.5 Conclusions

In this chapter, we explored two different methodologies for human facial recognition in image processing. We found that both had their own strengths, weaknesses, and intended purposes. We found that OpenCV was efficient and capable of locating subject faces in an image with almost no problems and few computing resources. We also found that DLib was excellent at locating facial features, at the cost of increasing the computational time to do so. This chapter has also demonstrated that a combination of the two was the most optimal way of producing the desired information. To meet the goals of this study, an additional level of logic was introduced to handle a moving, non-static subject and increase the reliability of identification. Finally, this work tested, verified, and demonstrated that these technologies can run efficiently on a low-power, low-cost hardware platform without significant loss of accuracy or performance.

Some additional functionality could be added in the form of the continued tracking of the subject. When a subject rotates their head about the x or y axis (the z axis is perpendicular to the plane of the image), the tracker is not designed to handle this change in geometry. Functionality to allow the identification of faces in profile or perspective would be a useful addition to this tracker. In addition to the increased tracking, once such an internal state was stored, the tracker would also be able to cut down on computational processing time, as it would only check a relatively small region of the input image, oriented to include the last known coordinates of the tracked face. This would be a minor improvement, but would affect a significant number of frames, and thus increase the frame rate of the program.

Bibliography

[1] OpenCV https://opencv.org

[2] Zou W *et al* 2009 Development of the embedded vision system: a survey *Meas. Control* **42** 216–21

[3] Viola P and Jones M 2001 Rapid object detection using a boosted cascade of simple features *Proc. 2001 IEEE Computer Society Conf. on Computer Vision and Pattern Recognition* CVPR 2001 *(Kauai, HI, USA)* p I

[4] OpenFace https://github.com/cmusatyalab/openface

[5] Torch https://torch.ch

[6] DLib http://dlib.net

[7] Rabhi Y, Mrabet M and Fnaiech F 2018 A facial expression controlled wheelchair for people with disabilities *Comput. Methods Programs Biomed.* **165** 89–105

[8] Pham T, Kim S, Lu Y, Jung S and Won C 2019 Facial action units-based image retrieval for facial expression recognition *IEEE Access.* **7** 5200–7

[9] Sajjad M, Nasir M, Ullah F, Muhammad K, Sangaiah A and Baik S 2019 Raspberry Pi assisted facial expression recognition framework for smart security in law-enforcement services *Inf. Sci.* **479** 416–31

[10] Cao C, Hou Q and Zhou K 2014 Displaced dynamic expression regression for real-time facial tracking and animation *ACM Trans. Graph.* **33** 43

Chapter 4

The roles of delay and power optimization techniques in VLSI design

Payal Bansal, Devendra Kumar Somwanshi, Balwinder Singh Dhaliwal and Pallavi Sapkale

This chapter describes different approaches used in the delay and power optimization of circuits. Various issues related to these approaches are also discussed. We include a detailed description of power dissipation in logic circuits and its minimization techniques, using low-power designs. The amount of dissipated power is defined by the rate of energy delivery from the source to the device/system. In battery-operated systems, the quantity of energy stored within the battery is restricted. Therefore, power dissipation is a vital metric of portable systems, because it defines the typical lifetime of the battery.

The chapter covers the development of new design methodologies that simultaneously reduce power, area, and delay. Power dissipation does not scale linearly with delay.

4.1 Delay optimization

In this section, we provide the reason why this chapter selects 'delayed performance in realistic circuits.' Integrated digital circuits are ubiquitous in modern life, and most of them are installed in mobile devices that have limited power sources (e.g. cell phones, clocks, and laptops).

The complementary metal–oxide–semiconductor (CMOS) process follows Moore's law, which states that 'the number of transistors that can be incorporated into a microcircuit will double almost every two years.'

Accurate delay is a requirement for speed. In efficiency studies, the clock delay is often considered to be a heavy and rapid value, regardless of the inclinations of the input and the output. There is also the option of linking a variable input to the same delay, ignoring the difference between the output's rise and fall times. These

doi:10.1088/978-0-7503-3919-3ch4

problems affect the speed of logic gates and should therefore be taken into account in estimating the delay.

4.2 Area optimization

The local model does not take into account the location of the roadway; this is acceptable at the transistor scale, but causes a major problem for the construction of realistic circuits using the variable input delay (VID) process. In order to achieve high durability and high performance, the sizes of CMOS technology features and threshold strengths have been declining for decades.

The area required is increased by the use of VID techniques, due to the use of external buffers, so this process could still be improved in terms of space reduction.

4.3 Variable input delay

In this chapter, we target the VID technique to achieve the reduction of delay. VID is a method for repairing clock signals. Clock delay contributes a major portion of the dynamic power use. In very synchronous circuits, many modules require a square-wave signal to be clocked at the same time. Hence, the unused clock signals consumed by different modules result in high power dissipation. Therefore, this strategy avoids the delay caused by providing clock signals at unused gates.

4.3.1 The concept of the variable input delay technique

VID saves power by adding more logic to a circuit to prune the clock tree. Pruning the clock disables portions of the circuitry so that the flip-flops in them do not need to switch states, because switching states consumes power. When flip-flops are not being switched, the switching power consumption goes to zero, and only leakage currents are incurred.

Figure 4.1 shows the concept of the VID technique, which uses a NAND gate. The VID method works by taking the 'enable' signals from the buffers, and applying them to the gate. It is thus necessary for the design to contain these enabling conditions to use and benefit from versatile input delays.

This method also extend the circuit's lifetime as a byproduct, however it increases the amount of buffers in the circuit. Improving power consumption has become a significant challenge in modern chip design, in particular, because of the limitations caused by leakage power requirements. The special methodology of the insertion of

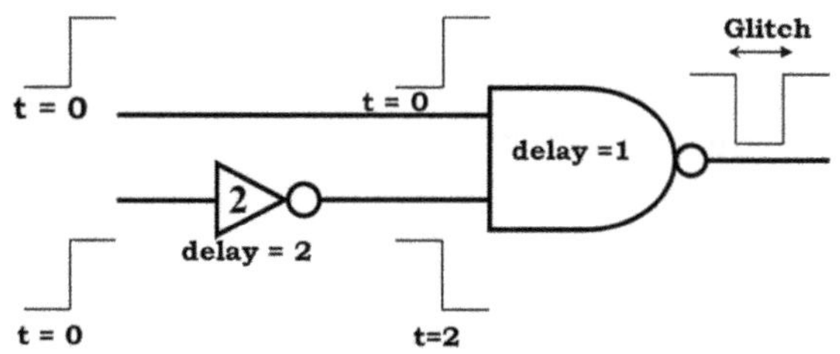

Figure 4.1. Block diagram of the VID concept.

decap cells is required to decrease the intensity of infrared (IR), thereby reducing the instantaneous current demand within the structure. Power consumption in circuits is usually documented, because it is the product of a variety of variables (Nt) and intermediate transformations (Pt) [7]. In this study, our main motivation is to reduce NGO by fault elimination in selected connected cells that contribute to high IR reduction. IR levels decrease over time and also adhesive cells gift thus area unit sometimes determined in dynamic IR analysis. Using the delayed input methodology mentioned above with a transmission gate as the compensation region for selected connecting cells, we can reduce the highest current contribution caused by glitches, so IR eye. The dynamic voltage (IR) decrease, compared to a vertical drop, depends on the performance of the setting, and so depends on the vector. A dynamic IR decrease requires an examination of the IR decrease caused by a broad range of circuit changes at the same time. One of the key requirements is to get a value change dump (VCD) file format that captures the transition information that causes important cell delays and connections created by the processing of the standard format delay (SDF) by gate-level simulation. Such a simulation captures the benefit of job modification by the planning of a slow window. Throughout IR analysis, the dynamics are separated into short measures. The length of a slow step is determined by a switch task window or intermediate change time that may be determined by a standing time analysis (figure 4.2).

Reducing $P(v, t)$ defects is the subject of this work. For this, we generally measure the square of the VID using a methodology that involves permanent insertion of a transmission gate at the inputs of computer circuits to reduce differing delays at the logic gate inputs. This approach can reduce defect breadth and therefore dynamic power.

4.4 System implementation

In this chapter, digital circuits are designed using the VID process. The chapter concludes with delays, power, and spatial creation in realistic circuits. We present a D flip-flop circuit and a full adder circuit that use the VID concept and compare the results with those obtained without the VID method. Both circuits are included in the VID process; the results are compared for both strategies and are presented in tabular form. This section examines the circuit description of the D flip-flop circuit and the full adder circuit. All parts of both circuits are also described in this section of the chapter.

4.4.1 D flip-flop

Flip-flop circuits are sequential circuits which a generate clock signal. They require feedback in the output-to-input path. Among flip-flops, the D flip-flop circuit is an easy and simple circuit which is used to create the clock circuits in CMOS technology (figure 4.3).

Figure 4.4 shows a block diagram of a flip-flop circuit which needs a feedback connection between the outputs and the input node.

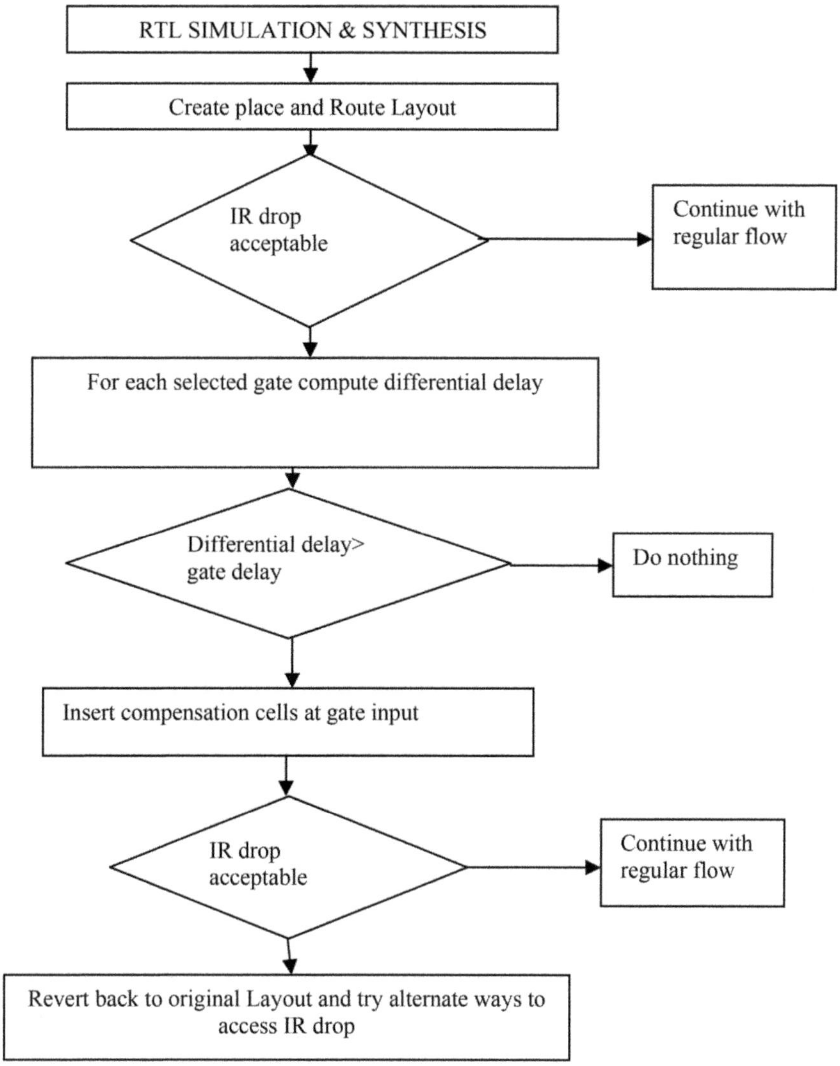

Figure 4.2. Flowchart of the required system.

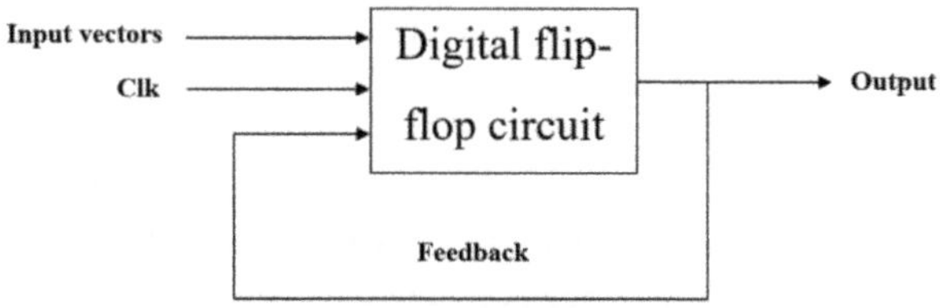

Figure 4.3. Block diagram of a flip-flop circuit.

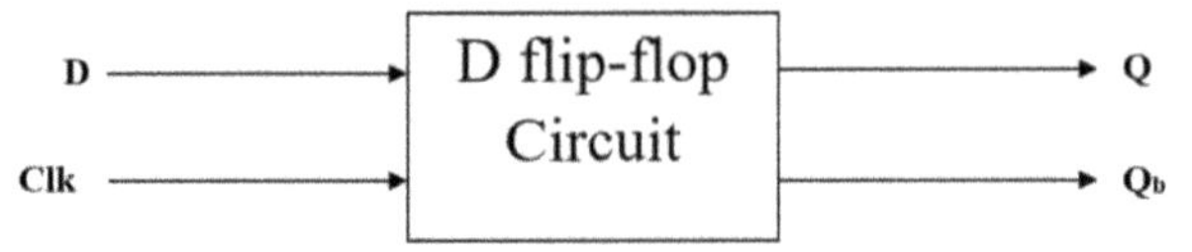

Figure 4.4. Block diagram of a D flip-flop circuit.

Table 4.1. Truth table of the D flip-flop circuit.

Inputs		Outputs	
CLK	D	Q	Q_b
0	0	0	16
0	16	16	0
16	0	0	16
16	16	16	0

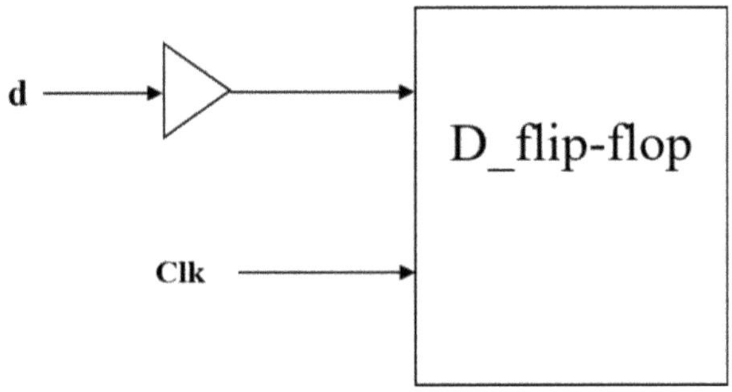

Figure 4.5. Implementation of a D flip-flop circuit using the VID technique.

A block diagram of the D flip-flop circuit is shown in figure 4.3, which shows that it has two inputs, D and Clk, at the input nodes of the circuit and two outputs, Q and Q_b, at the outputs of the circuit. The behavior of the circuit is defined by the truth table shown in table 4.1.

The implementation of a D flip-flop circuit using the VID technique is described below.

Figure 4.5 shows the implementation of a D flip-flop circuit using the VID technique. The extra switching activity of the clock is reduced by the buffers. The area is increased by this technique, due to the addition of the buffers.

4.4.2 Full adder circuit

The full adder circuit is known as a combinatorial circuit in the digital world. This circuit adds two 16-bit binary numbers and outputs two 16-bit binary numbers, i.e. the sum and carry values. A full adder is usually a component in a cascade of adders which add 8-, 16-, or 32-bit numbers or other bit widths of binary numbers.

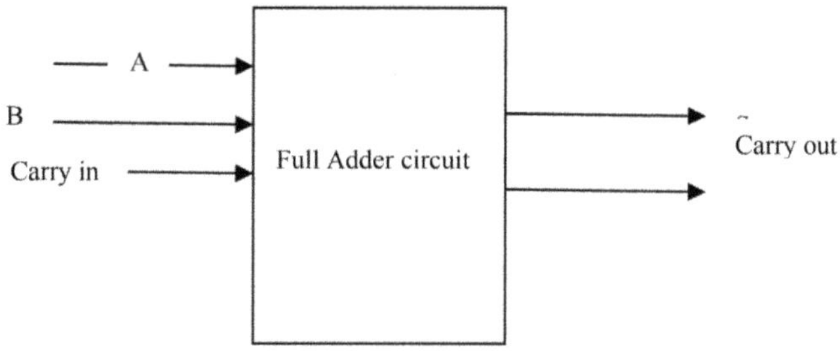

Figure 4.6. Block diagram of a full adder circuit.

Figure 4.6 shows the block diagram of a circuit which has two inputs, A and B, and two outputs, sum and carry. The circuit can be implemented with and without the VID technique. The extra switching activity of the clock is reduced by the buffers.

This section dealt with the theoretical aspects of the experiment. The technique is presented for different types of logic gate. The different types of parameter and minimization using the VID technique are also explained here. The next section presents the methodology and the steps that are involved in this project.

4.5 Digital Logic Design and Implementation

4.5.1 Architectural design of the work

In the nanometer regime, a large portion of the overall power consumption in high-performance digital circuits is caused by excessive currents. As a result of increased miniaturization and predefined power budgets, the power lost reduces the available power, impacting performance. Hence, techniques focusing on the area used are necessary to reduce the lost power while maintaining high performance. New low-power circuit techniques focused on the area used are needed to scale back the total current consumption in superior nanoscale circuits. The architectural design flow of the system design discussed in this chapter is shown in figure 4.7.

Figure 4.7 provides a conceptual description of the system design process, which uses Xilinx. The steps required for a system design in Xilinx are as follows:

1. Write the code for the system; in this section, we design the D flip-flop circuit. Here, the code is written using the Very High Speed Integrated Circuit (VHSIC) hardware description language (VHDL).
2. The simulation of the code is performed using the Xilinx-ISim simulator.
3. The resulting waveforms are displayed by ISim.
4. We compare the results with and without the use of the VID technique.

In the following, we discuss the flowchart of the VHDL simulation process for a D flip-flop circuit and explain all the steps required to design the circuit using the Xilinx-ISim simulator and compare the captured results using a synthesis report.

Figure 4.8 shows the design flowchart for the VHDL simulation of the D flip-flop.

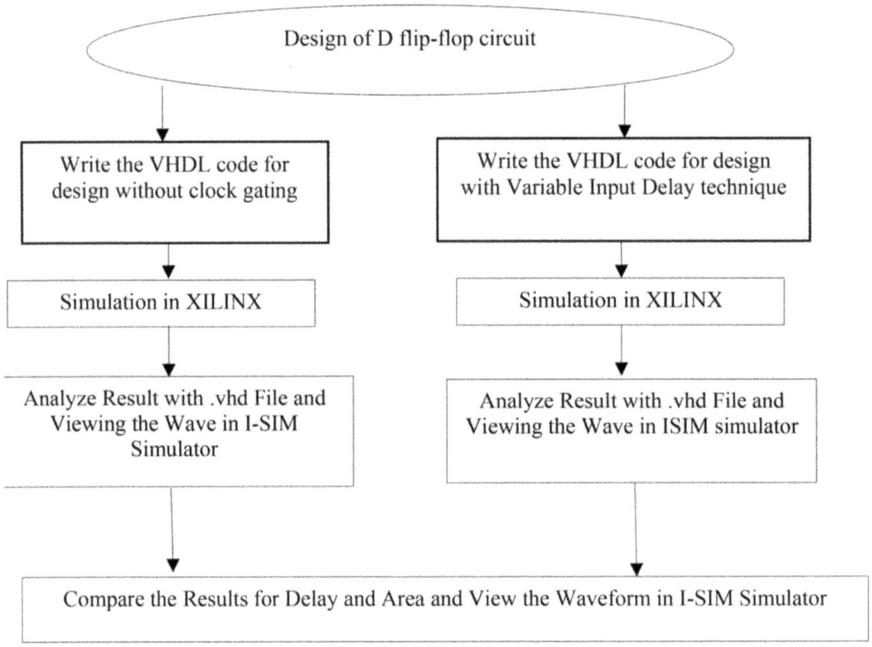

Figure 4.7. Architectural system design flow to address the delay and area of a D flip-flop.

First, we create a new project and select the VHDL as the top module. We then write the VHDL code for the D flip-flop circuit, first declaring an entity for the code that gives a description of all the input and output ports of the given circuit.

- D and Clk are given as the inputs of the circuit and Q, Q_b are given as the outputs of the circuit.

  We then define the architectural view of the circuit. In the architectural view, a behavioral and structural description of the circuit is given in Xilinx. In this section, a structural description follows for the circuit design, in which all the components are defined with their code and a behavioral description defining the behavior of the circuit corresponding to the truth table.
- The behavior of the circuit is as follows: if Clk and Clk event = '0', the circuit is off.
- If Clk and Clk event = '16', the value of input D is checked.
- If input D = '0', output Q = '0' and Q_b = '16'.
- If input D = '16', output Q = '16' and Q_b = '0'.

  We then synthesize the code. If the synthesis process is complete, we have completed the simulation process. If the synthesis process fails, we analyze the code for the circuit again.

  The simulation process is performed using the ISim simulator and defines the outputs for the input vectors '00,' '016,' '160,' and '1616.'
- We set clock and fall to '0' and rise to '16'; we set the value of D to '0' and '16.'

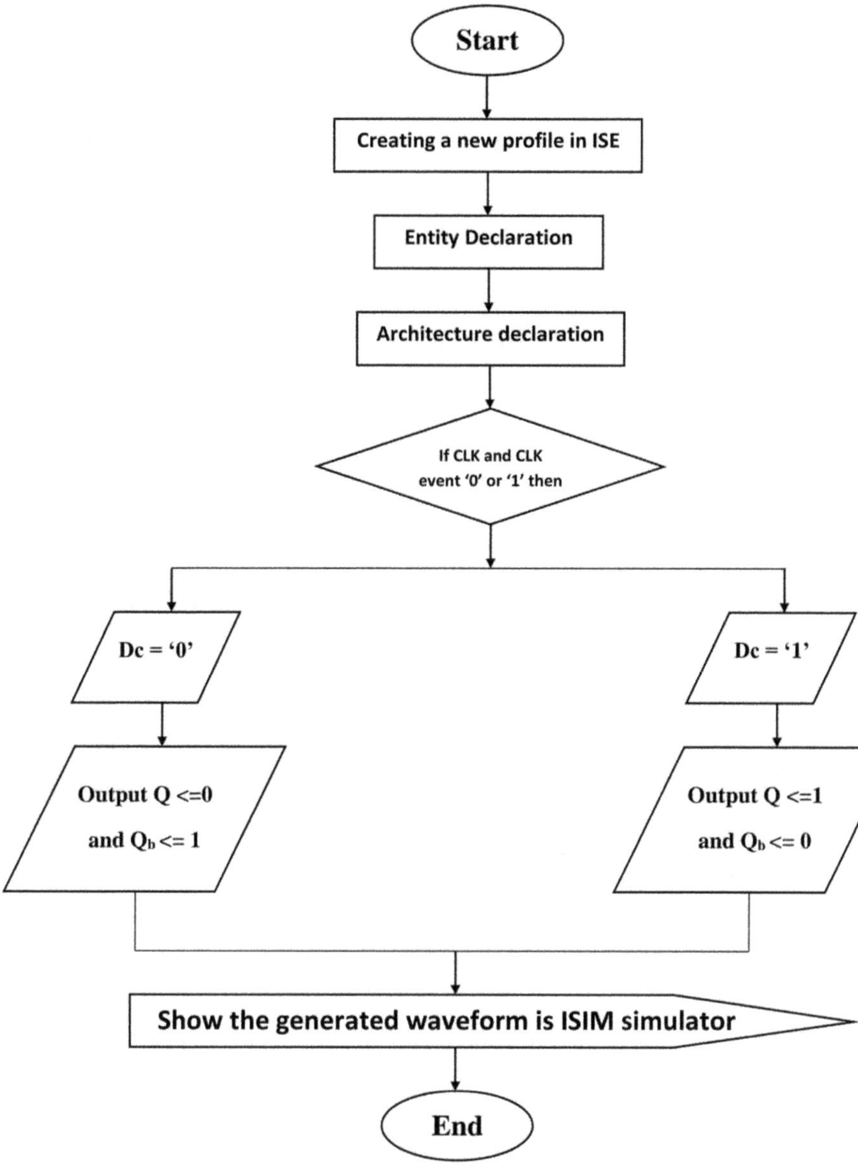

Figure 4.8. Flowchart of the VHDL simulation process for the D flip-flop.

- The output is generated for the corresponding input vectors. We check the waveforms against the truth table for the circuit.
- The simulated waveforms are displayed by the ISim simulator.

The flowchart of the delay and power optimization of a D flip-flop is shown in figure 4.9. (An explanation of the figure is given below.)

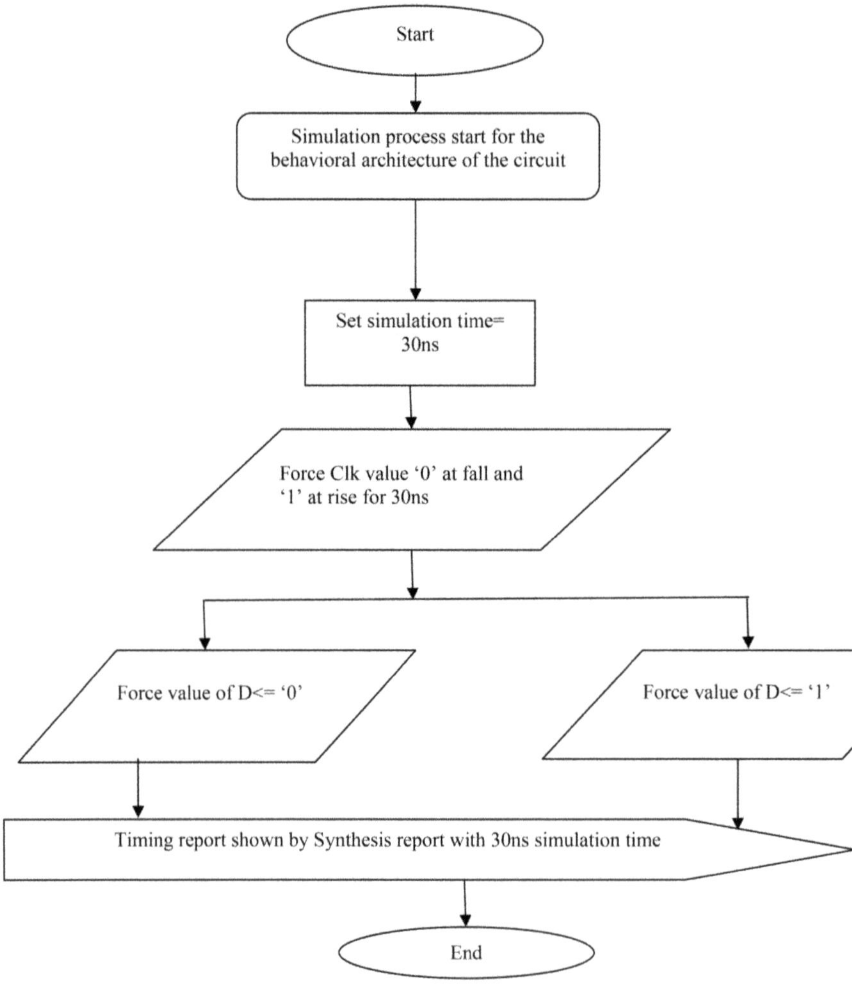

Figure 4.9. Flowchart of the synthesized code for a D flip-flop at 30 ns.

Figure 4.10 shows the flowchart for the simulation of delay optimization in logic circuits at a simulated time of 30 ns. From the figure, the delay is defined at 30 ns, the D inputs are set to '0' and '16,' and the clock inputs rise at '16' and fall at '0.' The timing report provides information about the delay of the circuit, which is given in the results. The delay results with and without the VID process are compared and given in tables 4.2 and 4.3.

4.5.1.1 Full adder circuit design

The design of a full adder circuit is implemented to optimize the delay in the circuit. The design flow of the delay synthesis process is shown below.

Figure 4.10 shows the design flow of a full adder circuit for delay and area synthesis. To design a full adder circuit using the VID technique, we first write the

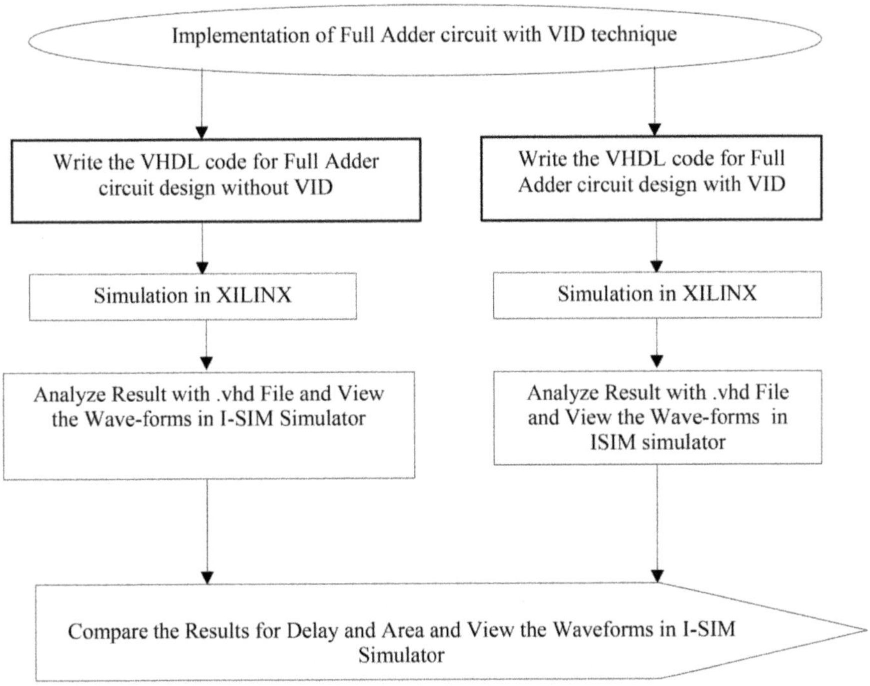

Figure 4.10. Architectural system design flow for delay and area synthesis in a full adder circuit.

Table 4.2. Simulation results for the full adder circuit without the use of the VID technique.

Components of the circuit	Delay parameters for the circuit without the use of the VID technique	
	Gate delay (ns)	Net delay (ns)
I_buffer	16.2168	0.622
LUT	0.704	0.420
O_buffer	3.272	
Total delay	6.236 ns	

Table 4.3. Simulation results for the full adder circuit with the use of the VID technique.

Components of the circuit	Delay parameters for the circuit with the use of the VID technique	
	Gate delay (ns)	Net delay (ns)
I_buffer	0.000	0.5165
LUT	0.704	0.420
O_buffer	3.272	
Total delay	4.161 616 ns	

VHDL code for the full adder circuit with and without the VID concept. We then simulate the code of the circuit, analyze the results in the .vhd file, and compare the simulation results and the synthesis report of the circuits with and without the VID concept.

The simulation is performed using the ISim simulator and the results are shown in table 4.3. The delay report and the area report are included in the synthesis report, which explores the improvement in the delay and the area.

4.6 Hardware/software used for experiments

In very large-scale integration (VLSI), different types of hardware can be used by the researcher for further experiments. In part two of this section, the designs are implemented on the Spartan 3E for asynchronous circuits (or any other types of circuit) to experiment with the data and obtain the results. The software packages used are:

- Xilinx-163.16 with the ISim simulator and VHDL structural modeling
- H-SPICE-2008.3

Software with a Cosmo-Scope wave generator was used to implement the circuit for power analysis. H-SPICE coding is used for the design of the system. The design flow with VHDL on Xilinx-ISE-163.16 is shown in figure 4.4 and an explanation of the design flowchart is given.

4.6.1 Introduction to the ISE design suite

This section includes a study of VHDL coding using the ISE software. VHDL is a high-level description language used to design digital circuits, and it also provides synthesis and simulation of the circuits. Xilinx-163.16 is also used, which provides a circuit optimisation improvement, compared to Xilinx-160.16. Xilinx-163.16 includes the ISim simulator, which provides circuit simulation. The ISE provides the register-transfer level (RTL) logic and test bench for the circuits, which produce a schematic for our design. It also performs the timing-driven synthesis and the physical synthesis. The timing-driven synthesis technique provides a significant improvement in design performance, and physical synthesis allows the design to be implemented using a net-list. In addition, the physical synthesis can reduce the number of cycles that are used in the implementation process.

To make it understandable and easy, we can use various functions, such as:

- Vertex 5
- Vertex 4
- Vertex III
- Vertex II
- Spartan 3A
- Spartan 3AN
- Spartan 3E
- Spartan 3

4.6.2 Features of the Spartan 3E

- Very low-cost, high-performance logic solution for high-volume, consumer-oriented applications
- Proven advanced 160-nanometer process technology
- Multi-voltage, multi-standard Selection™ interface pins
- Up to 376 I/O pins or 1656 differential signal pairs
- LVCMOS, LVTTL, HSTL, and SSTL single-ended signal standards
- Abundant, flexible logic resources
- Hierarchical select random-access memory (RAM) architecture

This chapter uses a structural description of the circuit for implementation.

- The behavioral description describes the behavior of the circuit, which is based on its truth table, algorithms, etc.
- The structural description describes the structure of the circuit; it defines all the components of the circuit. It is derived from any circuit equations and/or logic diagrams.
- On the basis of both descriptions, the code is analyzed in VHDL and a .vhd file is generated for simulation.
- The synthesis process is then performed using the .vhd file. The synthesized code generates an RTL and technology view of the circuit, which is shown by the RTL and technology schematic. The technology schematic provides a view of the circuit on a chip.
- We provide the input vectors for the input nodes of the circuit and perform gate-level simulation using the ISim simulator.
- The simulated results provide the output waveforms for the corresponding input vectors.

Thus, this section has provided the implementation structure of the circuit. It showed the design of the D flip-flop and the full adder circuit. The same circuit uses the VID technique for simulation purposes. The circuit is simulated using Xilinx and the waveform is analyzed using the ISim simulator. The next section focuses on the experimental details and provides the performance measurements of the circuit, such as the delay, power, and area of the circuit. It also presents the experimental analysis and simulation results.

4.7 Experimental results and analysis

As described earlier, Xilinx and H-SPICE are used to simulate the results of the described circuits. In this section, we discuss the details of the experimental analysis and the results obtained.

4.7.1 Experimental variations

In this section, we describe the design and simulation of two different types of circuit: a sequential circuit (the D flip-flop) and a combinatorial circuit (the full adder), while considering different design approaches, namely, those with and without the VID technique.

4.7.2 Experimental details

(a) Circuit description

We can design the logic circuit with a feedback circuit; in this case, the output depends on the history of the input elements. Sequential circuits have memory for storage purposes. The past input is encoded into a set of state variables using feedback; the state variables can be supplied with simple feedback or by the use of flip-flops. In this section, a D flip-flop is designed as a sequential circuit with and without the VID technique, and the delay is optimized in both circuits. We have already described the operation of the D flip-flop; here, we only consider the positive edge-triggered flip-flop (figure 4.11).

If the value of the clock is not at a positive edge, then the output value is retained. In other words, it does not change. A flip-flop additionally has two outputs, Q and Q_b. The output is the part that is kept the same; thus, the flip-flop usually outputs a little bit of information. A flip-flop may be a clocked device, which might modify its value only at positive clock edges. Once the clock is no longer at a positive edge, the flip-flop maintains (holds) its value. Flip-flops enable successive circuits to retain state (i.e., memory), which is a feature that combinable logic circuits do not have. The D flip-flop is taken as a secondary example in this section's work; the main target is the full adder circuit given below.

(b) Circuit description: full adder

The full adder circuit is a simple flip-flop circuit which is used to create circuits using CMOS technology. A block diagram of the full adder circuit is shown in figure 4.12.

Figure 4.13, shows the flowchart of the full adder circuit; there are four input parameters and sixteen output parameters in the circuit. The set of four inputs represented by A, B, C, and D are binary digits and output is calculated for all these inputs. The output set, Y, is also produced in the form of four binary digits.

This section considered the VID technique. The full adder circuit and the D flip-flop circuit were designed with and without the use of the VID concept and the results will be compared for both circuits.

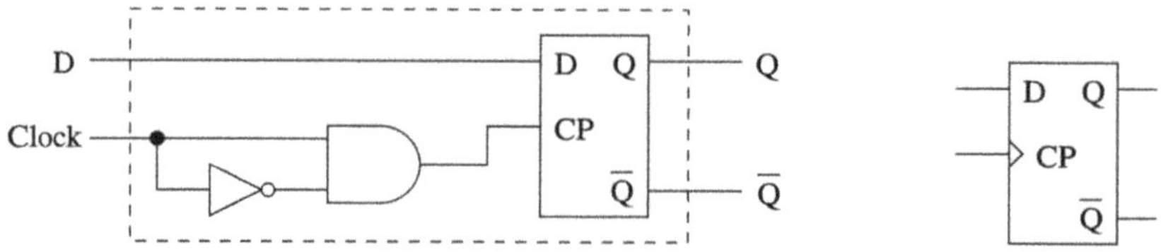

(a) Circuit diagram

(b) Logic symbol

Figure 4.11. D flip-flop circuit.

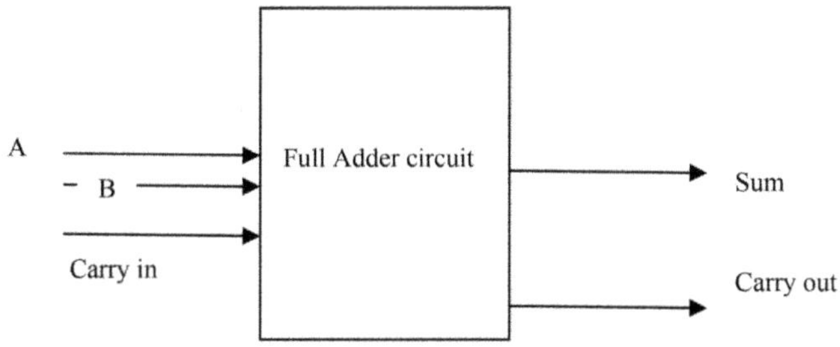

Figure 4.12. Block diagram of the full adder.

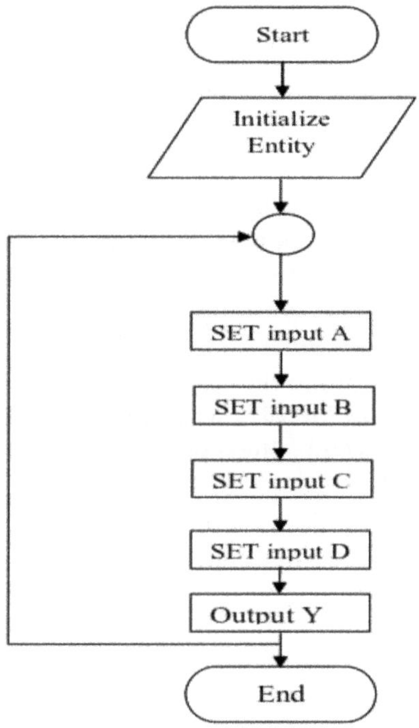

Figure 4.13. Flowchart of the full adder circuit.

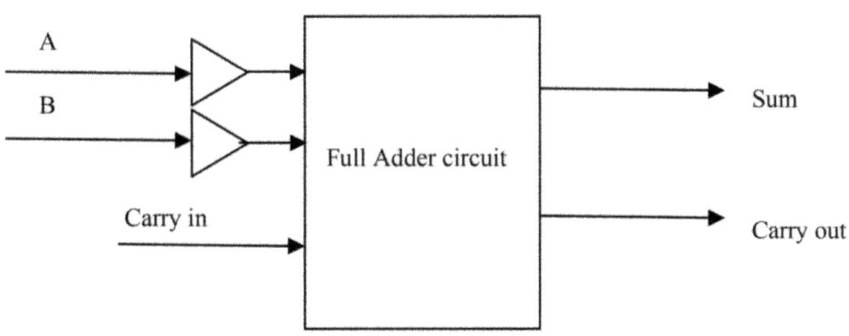

Figure 4.14. Full adder circuit using the VID technique.

The full adder circuit using the VID concept is shown in figure 4.14; buffers are added at the input nodes for delay reduction in the circuit. Using this concept, it is possible to reduce the delay in the circuit, as shown in the next section.

4.8 Results

In this section, the results of the chapter are presented in the form of a table showing delay and power for various combinations of the experiments carried out.

4.8.1 Simulation results for the full adder

This section considers the simulation results for the full adder circuit. The simulated work is shown in the figures below. The circuit was simulated using both techniques, i.e. with and without the use of the VID concept. A comparison of the results is shown in the tables in figures 4.15 and 4.16.

Figures 4.16 and 4.10 show the synthesis results for full adder circuits with and without the use of the VID concept, including the delay optimization of the circuit.

Tables 4.2 and 4.3 show the clock delay and net delay in a full adder circuit with and without the use of the VID technique. It can be seen that there is a delay reduction in the full adder circuit with the use of the VID technique.

The table shows that there is reduction in the clock delay and total delay when the VID technique is used, compared to the case without the use of VID. The logic delay for a full adder without the use of VID was found to be 6.236 ns whereas with the VID technique, it was 4.161616 ns; this represents a delay reduction in the circuit of 40% to 60% (see figure 4.17).

Figure 4.17 shows the simulated waveforms of the full adder circuit. The simulation results are valid for all input sequences, for example, 00, 016, 160,

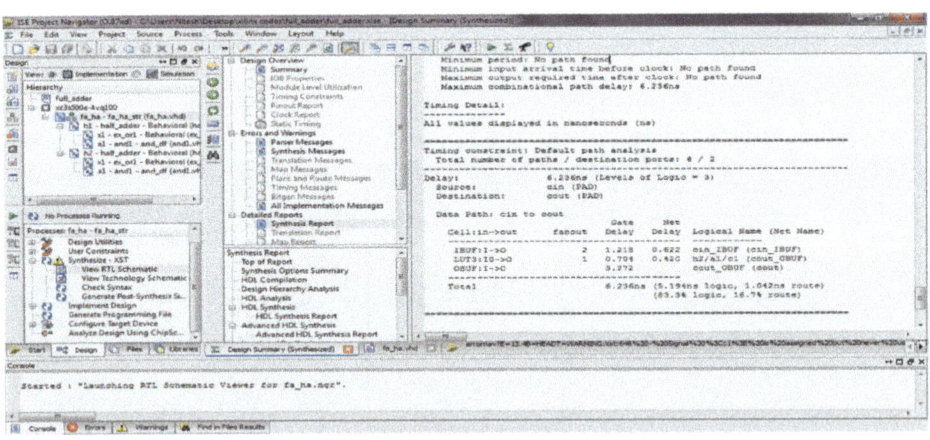

Figure 4.15. Synthesized results for a full adder circuit without the use of the VID concept.

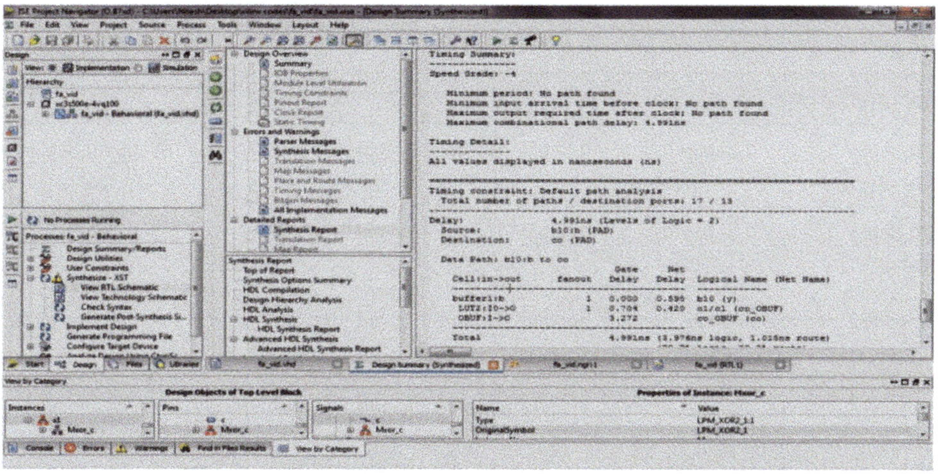

Figure 4.16. Synthesized results for a full adder circuit using the VID concept.

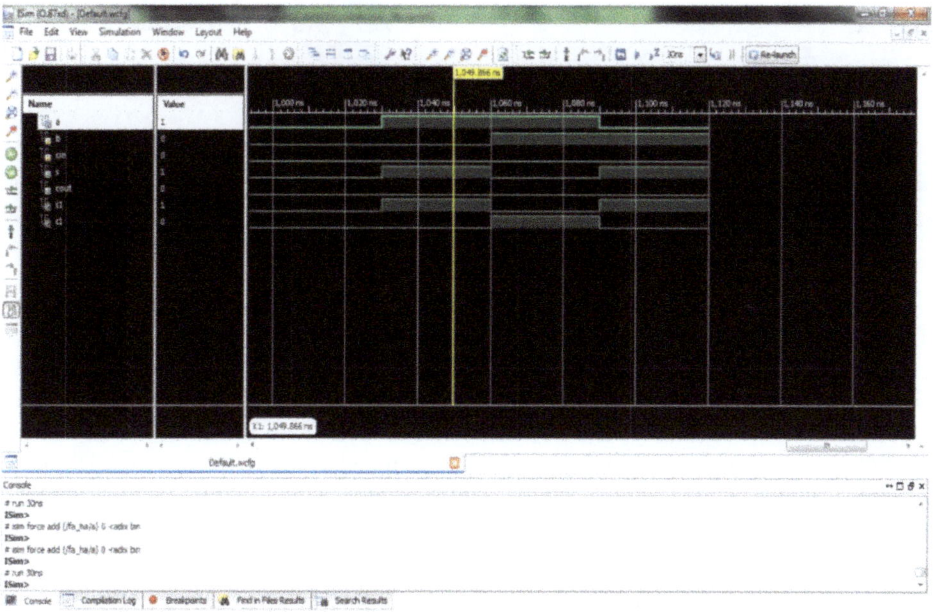

Figure 4.17. Simulation results for the full adder circuit.

1616 and the output sequences were as expected for their input sequences. The results were simulated for a predefined time period of 30 ns.

4.9 Comparative analysis of results

In this section, we provide a comparative analysis of the results in terms of delay. The results are presented in tabular and graphical forms. The comparative analysis is performed for results obtained with and without the use of the VID technique.

Table 4.4. Comparative analysis of the results for the full adder circuit with and without the use of the VID technique.

Circuit: full adder			
Delay parameters	Delay components	Without the use of the VID technique (ns)	With the use of the VID technique (ns)
Gate delay	I_buffer	16.2168	0.000
Net delay	I_buffer	0.622	0.5165
Logic delay	D flip-flop	6.236	4.161 616

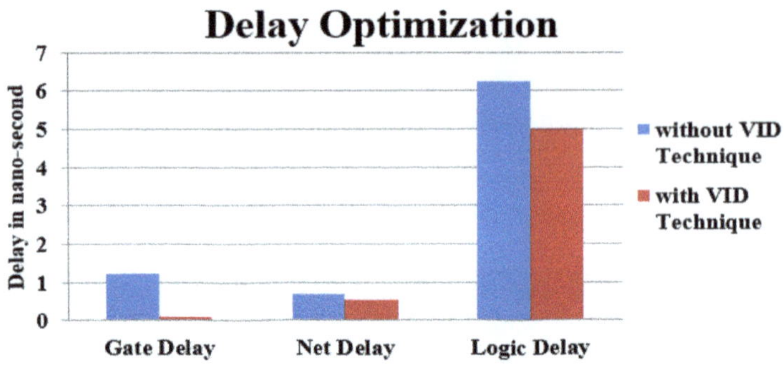

Figure 4.18. Comparative graph showing an analysis of the delay for a full adder circuit with and without the use of the VID technique.

In the comparative analysis of the full adder circuit with the VID technique, the gate delay, net delay, and logic delay parameters are reduced.

Table 4.4 shows the comparative analysis of the full adder circuit with the VID technique in nanoseconds; the gate delay is reduced from 16.2168 ns to 0.000 ns (a 1600% reduction); similarly, the net delay is reduced from 0.622 ns to 0.5165 ns (a 70%–80% reduction), and the logic delay is reduced from 6.236 ns to 4.161 616 ns (an 80% reduction).

A graphical chart of the comparative analysis of delay reduction using the VID technique in a full-adder circuit is shown below (figure 4.18).

Figure 4.18 shows a graphical presentation of the delay due to the full adder circuit. The chart shows that there is a 1600% reduction in the gate delay and a 70% to 80% net delay reduction with the VID technique. The total delay in the circuit improves by 80%. The delays are shown in nanoseconds.

4.10 Comparison of results

In this section, we perform a comparison between previous works that have used the 'VID technique' for different CMOS circuits and our results. In table 4.4, a total of three research works are compared with this study of the VID technique. For the full

Table 4.5. Comparison of previous work and the work described in this chapter.

Prior work				Presented work			
Circuit design	Technique used	Parameters	Optimized results	Circuit design	Technique used	Parameters	Optimized results
Two-input CMOS NAND gate	VID technique	Software: Spectre Analog Simulator and Cadence Technology: 0.25 μm CMOS technology, valid for all input sequences	Power optimized up to 58%. Overall delay was not increased.	Full adder	VID technique	Software: Xilinx Technology: 0.25 μm CMOS technology, valid for all input sequences	1600% reduction of gate delay, net delay reduced by 70%–80%, total delay improved by 80%.

adder circuit, the results for the delay compared with the VID technique show that a 40% to 80% delay reduction is possible with the use of the VID technique.

In earlier sections, we discussed the use of the VID technique in the design of full adder circuits, and, in this section, using a comparison of prior and presented work, we show that different circuits have been designed using the VID technique and compare the results obtained from different simulators. Below, we discuss the results for full adder circuits.

Table 4.5 shows the results of comparing prior work based on the VID technique and the results of this study. The prior work discussed a two-input CMOS NAND gate and used Analog Simulator and Cadence, while this study discussed the full adder circuit and used the Xilinx simulator. The table shows that reductions of 70% to 80% in the gate delay and the net delay are possible with the use of the VID technique, which is an increase from 50% to 52%, compared to the prior work.

As shown in the table, in the prior work, a two-input NAND gate was designed using the VID technique and a 52% delay optimization was performed using Analog Simulator and CADENCE. This is the same process as used in our study, in which a full-adder circuit was designed using the VID technique and a delay optimization of 70%–80% was achieved using the ISim simulator in Xilinx.

Thus, we have discussed the analysis of delay using by introducing the VID technique into sequential and arithmetic circuits. This chapter provided a comparative analysis of the VID technique. The area requirement was increased by the use of the VID technique, so this technique is not beneficial in terms of area optimization.

4.11 Conclusions

Continuous advancements in CMOS technology have enabled an exponential reduction of the space used by junction transistors and have allowed the switching speed to be scaled up over the decades. The reduction of gates through tapering is effective for superior energy reduction; however, this improvement quickly reaches the point of diminishing returns, causing a delay increment of half an hour or more, particularly when many hundreds of gates are connected to the output. Sadly, the circuit density has increased conspicuously as a consequence. Throughout this chapter, we have described various basic ideas associated with design improvement in VLSI circuits. The discussion has focused on various sources of power consumption in CMOS junction transistors; the key elements of power, such as dynamic, short-circuit, and escape power have been presented, followed by the fundamentals of the device logic-gate-level techniques used to scale back these power elements. Literature reviews associated with design improvement in logic circuits have identified gaps; for example, though much effort has been expended on power, delay, and space improvement, and researchers have worked on different application circuits and devised several strategies, the scope of this work has been restricted. In this chapter, we analyzed the behaviors of the D flip-flop circuit and the full adder circuit with and without the use of the VID technique. This method provides the most effective results in terms of delay, but consumes more space in the design process. The ISim package provided a visual representation of the waveform. We analyzed the behavior of the adder circuit with and without the use of the VID technique. The total delay due to the full adder circuit without the VID technique was 6.236 ns and with the use of VID in the full adder circuit, the result showed a delay of 4.161616 ns. A delay reduction of 70%–80% can be achieved, however the space used is increased by the external buffers.

Bibliography and further reading

[1] Gupta A 2016 Carry save adder in Boolean networks *IEEE Trans. CAD Integrated Circuits Systems* **25** 743–55

[2] Kder A May 2005 Logical effort method *IEEE Int. Symp. Circuits Syst.*

[3] Johri A, Schimpfle and Nassek J A 2016 MPSO algorithm for delay optimization *IEEE Int. Symp. Circuits Syst.* p 16

[4] Kabbani A, Al-Khalili D and Al-Khalili A J 2005 Delay analysis of CMOS gates using modified logical effort model *IEEE Trans. Comput.-Aided Des. Integr. Circuits Syst.* **24** 937–47

[5] Mishchenko A *et al* 2006 Using simulation and satisfy-ability to compute flexibilities in Boolean networks *IEEE Trans. CAD Integrated Circuits Systems* **25** 743–55

[6] Datta A *et al* 2000 Analytic models to estimate yield for a pipelined design based on delay distributions of individual pipe stages *IEEE Trans. Image Process.* **16** 1673–80

[7] Wroblewski A, Schimpfle C V and Nassek J A 2000 Automated transistor sizing algorithm for minimizing spurious switching activities in CMOS circuits *IEEE Int. Symp. Circuits Syst.* **3** 291–4

[8] Calhoun B H, Wang A and Chandrakasan A 2005 Modeling and sizing for minimum energy operation in subthreshold circuits *IEEE J. Solid-State Circuits* **40** 1778–86

[9] Girod B 2000 Efficiency analysis of multi-hypothesis motion—compensated prediction for video coding *IEEE Trans. Image Process.* **9** 173–83

[10] Lasbouygues B *et al* 2006 Logical effort model extension to propagation delay representation *IEEE Trans. Comput.-Aided Des. Integr. Circuits Syst.* **25** 1677–84

[11] Alpert C J, Devgan A and Quay S T 1999 Buffer insertion for noise and delay optimization *IEEE Trans. Comput.-Aided Des. Integr. Circuits Syst.* **18** 1633–45

[12] Sutherland E, Sproull B and Harris D 1999 *Logical Effort: Designing Fast CMOS Circuits (The Morgan Kaufmann Series in Computer Architecture and Design)* (San Francisco, CA: Morgan Kaufmann)

[13] Faria G, Henriksson J A, Stare E and Talmola P 2006 DVB-H: digital broadcast services to handheld devices *Proc. IEEE* **94** 194–209

[14] Glancer. Smith 2007 Automation design of high performance digital integrated circuits *IBM J. Res. Develop.* **26** 1606–6166

[15] Hashimoto 2002 *Circuit Delay Optimization* 3rd ed. (Pearson Education)

[16] Handrik and Walker W W 2000 *CMOS VLSI Design: Integrate Logic Synthesis and Physical Design* 3rd ed. (Upper Saddle River, NJ: Addison Wesley)

[17] Linholm L W 1975 An optimized output stage for MOS integrated circuits *IEEE J. Solid-State Circuits* **10** 106–9

[18] Jeffery J K 2000 Proposed distributed stream-based applications *Proc. IEEE Int. Symp. on Efficient Embedded Computing* pp 3220–3

[19] Tsitsikils J, Chow W-K and Sham C-W 2002 The complex resource and time coordination *IEEE Trans. VLSI Syst.* vol 20, *Engineering* vol 16

[20] Karthik and Wright P E 2006 Structural representation and technology mapping of a Boolean function in the design of non-regenerative digital logic circuits *SIAM J. Optim.* **16** 1647–16162

[21] Venkat K and Prasad S 2016 Arbitrary logic gates as buffers *Electrical Technical Reports* Purdue Libraries 16

[22] Kwang and Salankar S S May 2003 Logical topology design algorithm using local optimization technique *Proc. Annual IEEE India Conf. (INDICON)* pp 4–16

Chapter 5

SIVAS: smart interactive virtual assistance system—a voice user interface

Abhishek Bharti, Amardeep Kumar, Manonita Verma and Lochana Perera
Menka Yadav

The purpose of this chapter is to describe the implementation of the Smart Interactive Virtual Assistance System (SIVAS)—a voice user interface (VUI) application. It can effectively and efficiently translate human voice commands into computer actions. It can also act as a virtual assistant and can perform various tasks for the user, such as reading an article from Wikipedia, finding out the meaning of an unknown word, controlling the keyboard/mouse of the computer or other domestic appliances, etc.

Less able people, such as those with limited dexterity or vision impairment, constitute a large part of the total population. They are a very important part of our society, but they do not receive the same opportunities as others in terms of their inclusion in society. One of the important reasons for this is that they are unable to interact with technology, as everything is just a form of technology nowadays. Therefore, it is necessary to develop an easily accessible, reliable, efficient, and financially affordable VUI, so that they can also interact with technologies just by using their voices, thereby eliminating the effects of their disabilities.

5.1 Introduction

VUIs may be hardware devices or software applications, and are used to bridge the gap between users and machines. Users and machines have their own languages that they use and understand, but in order to communicate with each other effectively and efficiently, they need an interface that can act as a language translator between them. Therefore, a VUI acts as an interface between them and uses the human voice as the medium of communication.

doi:10.1088/978-0-7503-3919-3ch5

5.1.1 Aim

The aim of this chapter is to describe the design of a VUI application for low-end devices, which can help a wide variety of users by enhancing their control experiences.

5.1.2 Objectives

The main objectives of this application are as follows:

(a) Helping users with limited dexterity or vision impairments by removing the need for a keyboard, mouse, or touch screen to control their computers
(b) Enhancing the control experience and decreasing the effort required by automating most of the frequent and common computational tasks, providing an handy voice-based assistant
(c) Providing easy-to-learn, simple, but powerful commands to control each and every aspect of the system
(d) Development practices must adhere to the principles of simplicity, readability, modularity, layering, clarity, and efficiency of the design, architecture, and code.

5.1.3 Challenges

The paradigms of VUI design pose a large number of challenges to usability. As practices for VUI designs are still emerging, while designing SIVAS, we encountered various essential challenges and issues, which are described below:

(a) **Transcription**: VUIs allow their users to provide speech input, but directly analyzing users' speech is very difficult due to large variations in the properties of speech data. Therefore, we need to perform speech-to-text (STT) conversion using speech recognition (SR) technology, so that the system can work on text data (which is more invariant, compared to speech data) to understand users' queries.
(b) **Activation**: most systems today still rely to some extent on non-speech input to activate the VUI. However, the VUI must be continuously listening and rely on a speech-based mechanism for its activation. Therefore, we use a wake-up word (WUW), which is a special word or phrase that is meant to activate a device when spoken. There are a few famous wake-up phrases such as 'Hey Siri', 'Ok Google' and 'Alexa'. The WUW must be spoken in the context of requesting attention. After the successful detection of the WUW and its alerting context, the VUI may safely interpret the following utterance as a query. A good choice of WUW can also substantially increase the accuracy of the VUI. Therefore, a WUW should be unique and promote the brand of the application using it. The WUW should be a medium-sized phrase and contain a variety of sounds.
(c) **Discoverability**: in order to understand the scope of VUI capabilities, a VUI needs to provide easy, short, relevant, memorable, and flexible commands

along with a cheat-sheet containing all the commands with working examples for ease of reference.

(d) **Understanding**: designing an effective system in order to maximize conversation-based understanding is still an open area of research. VUIs that can interpret and manage conversational state are challenging to design, due to the inherent difficulty of integrating complex natural language processing (NLP) tasks, such as skill detection, intent classification, named identity recognition, slot filling, information retrieval, coreference resolution, and dialog management. Nowadays, most voice assistants can easily execute a single command but may face some difficulty when faced with complexity or may be limited in their ability to detect the specificity of commands used for fine control of the system, such as those used to control the mouse and keyboard of the computer. In order to support fine control of the system, we have provided a 'cursor mode.' This mode is used to provide features for controlling the keyboard/mouse of the system, explicitly and exclusively.

(e) **Presentation**: describes how the outputs are delivered to the users in a way that enhances productivity by providing ease of reading, listening, and understanding and analyzing the output. While presenting the output to the user, we should focus on catching the user's attention, regardless of their background, age, gender, or language. Generally, this includes three essential aspects, namely layout, color, and font.

5.1.4 Use cases

The enormous capabilities of VUIs can be explored for a broad range of application areas, from daily lifestyle to advanced fields of science and technology. Some of the interesting and promising use cases are mentioned below:

- They can provide assistance to people who have problems such as limited dexterity and vision impairments, so that they can control machines that they encounter in daily life, such as the computer, cell phone, TV, etc. just by using their voices, without any fear of shock or knowledge of the technicalities.
- They can be helpful in allowing users avoid touching public surfaces during pandemics. Users can use their voices to open office doors, press elevator buttons, control public screens, etc.
- They can remove the requirement for a screen, touch pad, keyboard, mouse, etc. and allow users to communicate with their devices just by using their voices. This helps them to reduce their screen time and allows for a hands-free, eyes-free experience, in which users can interact with their devices while focusing their attention elsewhere.
- They can be used as a default interface for commanding self-driving cars.
- They can be very useful in emergency situations, in which the user cannot move and wants to contact the emergency authorities and services, as the user can use their voice to call the police, ambulance, fire department, etc.

5.2 Literature review

Recently, people have started using **'Portable Embedded Systems'** on a very large scale, and due to this, our lives have started becoming heavily dependent on these systems, as the Internet of Things (IoT) is gaining in popularity.

These systems are nothing more than an optimized integration of hardware and software, which together can be seen as a single component or electronic device that can be used by people in their daily lives, in order to make it easier to live. They are generally very limited in their functionality and mostly used for specific processing, especially in real-time systems, because of their low latency and ability to meet critical deadlines.

They generally consist of microcontrollers, which contain the microprocessor, memory for storing programs and data, and analog-to-digital and digital-to-analog converters, such as microphones, speakers, actuators, sensors, etc.

Recently, these systems have started to be used in a wide variety of fields, such as automobiles, railways, aircraft, aerospace, medical, home, clothing, factories, etc. Our society is being impacted by these applications, in terms of living, health, security, and privacy. Embedded systems now account for around 98% of all processors being installed, and most of them are not even visible to their end users as computers. This means that they are almost omnipresent and ubiquitous in nature, and they also form the basis of a significant economic push.

Embedded systems are available to meet an enormous number of use cases in almost every field of science; in particular, they are becoming an integral part of the biomedical field. In the past, people generally used these systems to take care of issues that demanded very frequent monitoring and maintenance. Today, they also function as aids that help less able people, in order to eliminate or reduce the effect of their disabilities in their day-to-day lives and bring them closer to modern technology.

Currently, there are various microcontrollers or similar portable processing hardware devices available on the market, such as the Raspberry Pi, the Arduino, ESP32, STM32, etc. All of these provide development boards with the different required configurations, which can be used to run various programs, generally written in C, C++, Python, etc. They can also interact with the outside world using the wide variety of I/O sensors available, which allow them to be extensively used in/as embedded systems.

Voice assistants, such as Amazon Echo, Apple Siri, Google Home, and other voice-controlled smart devices have opened up a lot of opportunities for implementing VUI services on portable embedded systems that can be integrated by the healthcare industry, in order to make machines voice enabled and increase their controllability. In online browsing, more than 30% of current access takes place via voice searches, and the use of voice-assistant apps for multiple purposes is increasing at a remarkable rate. By the end of 2017, voice-based e-commerce or shopping using voice had reached over $2 billion in value, and is expected to reach $40 billion by 2022.

5.3 Methodology

The architecture of this application is divided into two layers, namely the *front-end* layer and the *back-end* layer, as shown in figure 5.1. The front-end layer runs on a JavaScript-based process, while the back-end layer runs on a Python-based process. Both processes run concurrently and are responsible for performing specific tasks. They both connect with each other using a client–server model, in which the front-end layer acts as the client and the back-end layer acts as the server. As they use the client–server model, they therefore perform interprocess communication (IPC) using the WebSocket communication protocol (Request for Comments (RFC) 6455), which provides a full-duplex communication medium over a single transmission control protocol (TCP) connection.

The core reason for the division of the application's architecture into layers is to provide the flexibility to run different layers on different systems. This flexibility helps users to scale up the application by moving the back-end layer (server) from local devices to the cloud, where they can reside and connect with the front-end layer (client) via the same WebSocket IPC as before. In short, if local devices can handle and support both layers, then both layers can run on the local devices, but if the local devices cannot handle or support them, then only the front-end layer needs to run on the local device, while connecting to and making use of the back-end layer, which is available in the cloud. This layered architecture allows users to change the distribution of computing resources between centralization and decentralization, according to the needs of the situation. This approach can be very useful when deploying this application in the IoT, in which local devices have very few computing resources (in terms of both power and memory), but they can perform their processing in the cloud.

The front-end layer is composed of two functional units, namely the *input* and *output* units. The back-end layer is also composed of two functional units, namely, the *processing* and *retrieval* units, as shown in figure 5.2.

5.3.1 Initialization

Before doing anything else, it is very important to run both processes and connect them to each other using the WebSocket mechanism. This requires an initialization phase, as follows:

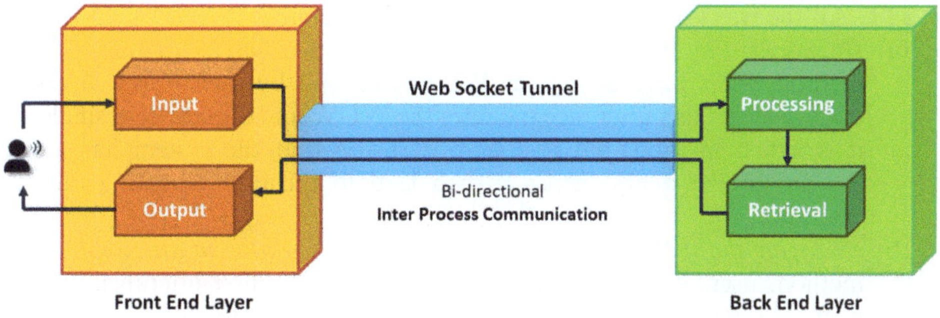

Figure 5.1. Layered architecture.

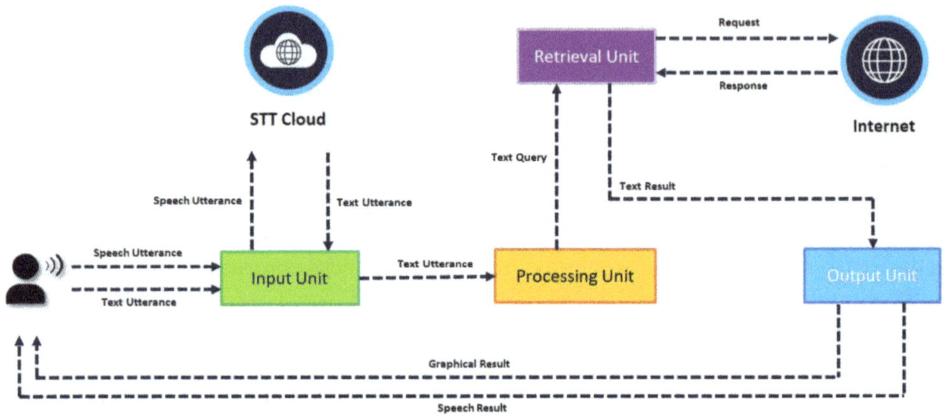

Figure 5.2. The high-level data flow cycle.

(a) Start the execution of the Python-based process.
(b) This process creates a TCP server on the local host at port 12000 and waits for a TCP client to connect to it.
(c) We can now start the JavaScript-based process by accessing URL 'http://localhost:12000' using a web browser.
(d) The above step automatically starts executing a JavaScript-based process which creates a TCP client and connects with the TCP server (running using the Python-based process).
(e) Once the TCP client is connected to the TCP server, they both upgrade their connections and automatically create a WebSocket tunnel between them for IPC.
(f) Both layers (processes) are now running concurrently and are also connected to each other. At this point, the application is ready to perform its functions.

5.3.2 Input unit

This unit is responsible for taking input from the users in various ways. This unit is a part of the front-end layer, because it uses the speech recognition API of JavaScript for STT. All the ways of providing input are listed below:

(a) *By saying the WUW followed by a query in a single utterance:* In this method, users have to say the WUW followed by their query in a single utterance. As this utterance includes the WUW and a query in a single step, it is a fast method, but it does not ensure that the system has actively listened to the query; hence, it can lead to the user repeating the same utterance.
(b) *By saying the WUW and then saying the query in different utterances:* In this method, users only have to say the WUW in the first utterance and then wait for the system to activate. Once the system is activated, users can then say their query in the second utterance. As this includes two steps, it is a

slow method, but it does ensure that the system is actively listening to the query.

(c) *By saying the query directly without the WUW:* In this method, users only have to say the query; hence, this method does not require the user to say the WUW. This method is specifically applicable and used when the application runs in cursor mode. As it does not require the WUW to be used in the utterance, it removes the overhead of the WUW; it the fastest method and increases the usability of cursor mode, in which users have to say a sequence of commands frequently.

(d) *By tapping the activator button and then saying the query directly without the WUW:* In this method, users have to press an activator button first to activate the system. Once the system is activated, users can say their query in a single utterance without the WUW. As this includes two steps, it is a slow method, but it does ensure that the system is actively listening to the query.

(e) *By typing the query directly into the query box:* In this method, users can type their queries into the query box. This is useful in handling situations such as those when the SR cloud is not working, the user is offline, or when noise is present near the microphone.

If the input query is in the form of text, then the input unit directly passes it to the processing unit, but if it is in the form of speech, the input unit first converts it into a text query using STT technology and then passes the converted text query to the processing unit for further processing.

5.3.3 Processing unit

This unit receives the text query from the input unit, and is responsible for sequentially applying various NLP techniques to the user's text query, which convert the user's query into commands so that the application can understand what actions it needs to perform in order to fulfill the user's query. This unit is a part of the back-end layer because (i) Python has rich and advanced NLP libraries which can be used to process and understand the query, and (ii) as the processing unit may require a large amount of computing resources, it can easily move to the cloud whenever required, as it is a part of the back-end layer.

In order to process the user's query, it performs the following steps in sequence:

(a) **Text preprocessing:** this step performs the following operations on the query string:

It removes extra white spaces from the beginning and the end of the query string.

It converts the query string into lowercase characters.

It performs *tokenization* on the query string. This converts the string into a set of words so that each word can be easily analyzed separately, along with its position in the query.

It performs *lemmatization* on the set of words, which reduce the words to their base forms, hence standardizing words with similar meanings to their root.

(b) **Skill detection:** this step is used to establish whether the query string contains any predefined skill name or not. In order to check this, the query string is compared with two given regular expressions (RegEx) templates.
ask (skill), (query)
(query) on (skill)

The first RegEx allows users to provide the skill name at the beginning of the utterance using the mandatory word 'ask.' For example: *Ask dictionary to define quarantine*. In this utterance, skill = 'dictionary' and query = 'to define quarantine.' However, the second RegEx allows users to provide the skill name at the end of the utterance using the mandatory word 'on.' For example: *Search meaning of quarantine in dictionary*. In this utterance, skill = 'dictionary' and query = 'search meaning of quarantine.'

As we can see, users can include a skill name in the utterance in either of the abovementioned ways. Both ways are equally functional and important; the choice depends on the user, who can use one approach for some requests and the other for the rest of their utterances.

If no skill name is found in the utterance then the whole utterance becomes the query and the 'default' skill is assigned automatically. For example: *Tell me some news*. The results of this utterance are: skill = 'default' and query = 'tell me some news.'

(c) **Intent classification:** once the skill is detected or assigned, the next task is to identify the intention of the user within the scope of that skill. Generally, there are two types of user intention.
Intentions that involve entities
Intentions that do not involve an entity

Intentions that involve entities: in this case, one or more entities can be present in the query. These entities are like parameters that can fill the empty slots and complete the command. In order to handle this, we use a template-matching technique, which compares the query with predefined templates for different intentions to try to extract and recognize the entities. This is used to extract the variable words (variable length entities) from the query using RegEx.

We now explain this process in detail using an example based the 'dictionary' skill, which illustrates two intentions that use one entity.

- Intention-1: 'show meanings of word'
  Template-1: search meanings of [word]
  Template-2: define [word]
- Intention-2: 'get antonyms of word'
  Template-1: search antonyms of [word]
  Template-2: antonym for [word]

Intentions that do not involve an entity: in this case, no entity is present in the query. The query only contains important and specific keywords, irrespective of their position in the query. In order to classify this intention, we create a hierarchy of bags of keywords and try to find the intersection between the bags and the query at each level. This technique is used to classify the intention using a time complexity of $O(\log n)$ instead of $O(n)$. We have to check the whether the query contains a primary set of words or not; if not, we keep checking further depths of the hierarchy, so that at last, we end up with a required action that needs to be performed.

5.3.4 Retrieval unit

This unit receives input from the processing unit, which tells it about the information the user wants. Therefore, this unit is responsible for retrieving the required information from local storage or from the internet, which will be used to fulfill the user's demand. In order to retrieve the information from local storage, it directly opens the required file(s) using their file system path(s) and delivers the information accordingly. But in the case of retrieving the information from the internet, it uses two popular mechanisms, namely *REST APIs* and *web scraping*.

The representational state transfer (REST) application programming interface (API) is a developer-oriented mechanism that provides data to users according to set of constraints defined by the developers or the data source. The client can request data from the API by sending hypertext transfer protocol (HTTP) requests and receives the data via HTTP responses.

On the other hand, web scraping is the process of opening a website, capturing its content, and using that content to get the required information.

5.3.5 Output unit

This unit receives query results in text format from the retrieval unit. It is responsible for converting text results into graphical and speech outputs using predefined HTML/CSS templates and text-to-speech (TTS) technology, respectively, so that users can see/read results on the screen and can listen to results through the speakers. It is a part of the front-end layer because it uses the JavaScript Web Speech API to perform TTS operations and its HTML/CSS-based graphical outputs can be displayed in an ordinary browser window.

5.4 Functionalities

This application provides two modes of operation, namely *command mode* and *cursor mode*. Each mode has its respective and distinct features, functionalities, and usability.

5.4.1 Command mode

This mode requires the user to provide the WUW with the query every time, in order to activate the system. It provides automated skill functionality and can execute predefined sets of commands together to perform specific tasks. This application provides one default and four third-party skills, namely YouTube, dictionary, Wikipedia, and IMDb.

5.4.2 Cursor mode

This mode does not require the user to provide the WUW with the query, as this mode always automatically remains in an activated state. It provides the full functionality of the keyboard and mouse, so that the user can produce mouse clicks and key presses via voice only. This can be very helpful in tackling novel situations, such as handling captchas, pop-up windows, etc.

5.5 Hardware implementation

5.5.1 Introduction

This application can run on portable, standalone, and low-end hardware devices, which makes it usable in the field of the IoT. The hardware of this system model is designed so that the main component is the Raspberry Pi, which acts as a processor for the whole system and is also connected to various other peripherals, which act as input/output (I/O) terminals; see figure 5.3.

5.5.2 Required devices

(a) Raspberry Pi 3+
(b) 5 V DC power supply
(c) HDMI LCD
(d) USB microphone
(e) Speaker with a 3.5 mm jack (figure 5.3)

Figure 5.3. Block diagram for hardware implementation.

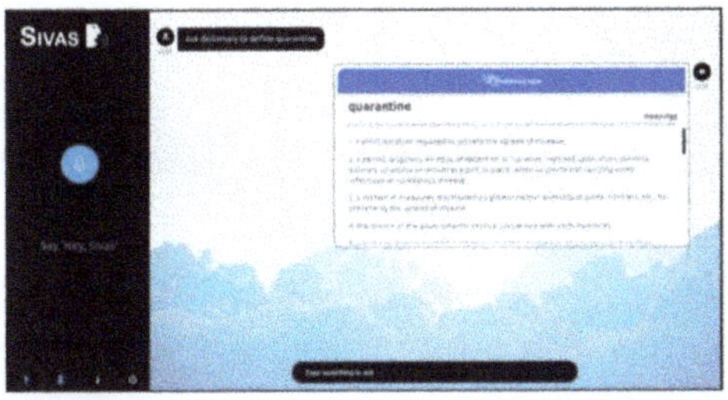

(a) Dictionary - Show meanings for given word

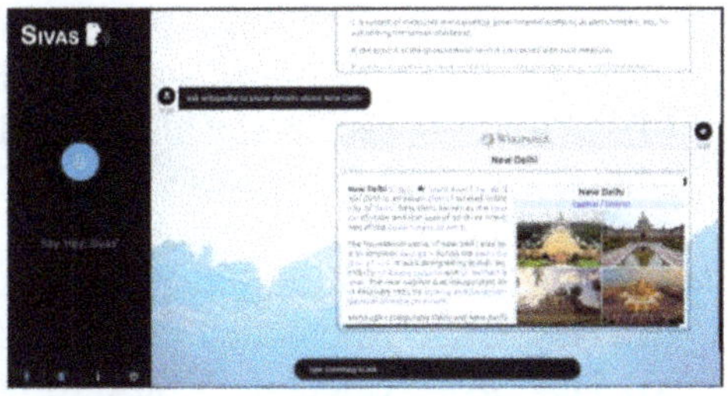

(b) Wikipedia - Show details for given title

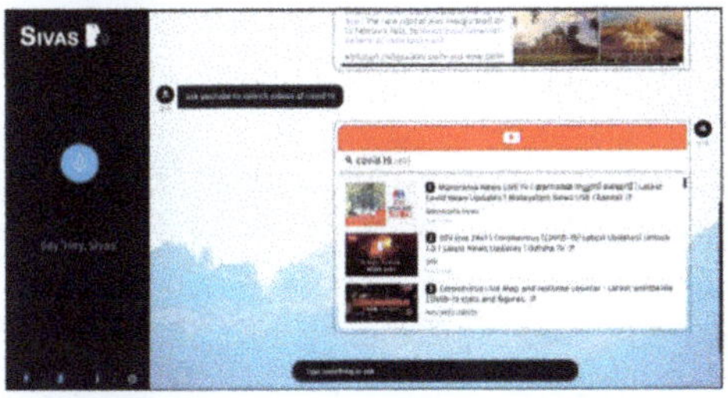

(c) YouTube - Search videos for given title

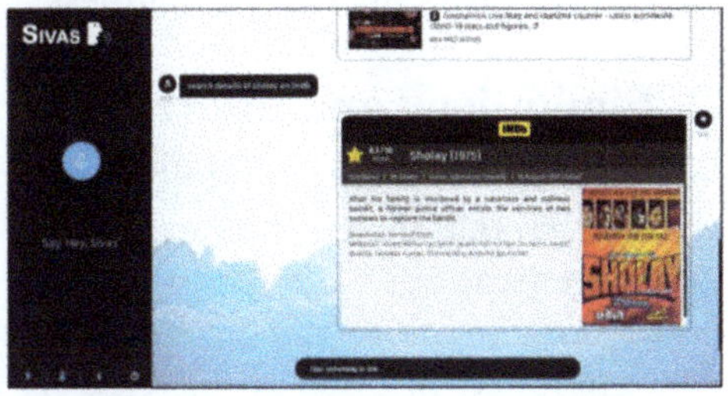

(d) IMDb - Show details for given movie

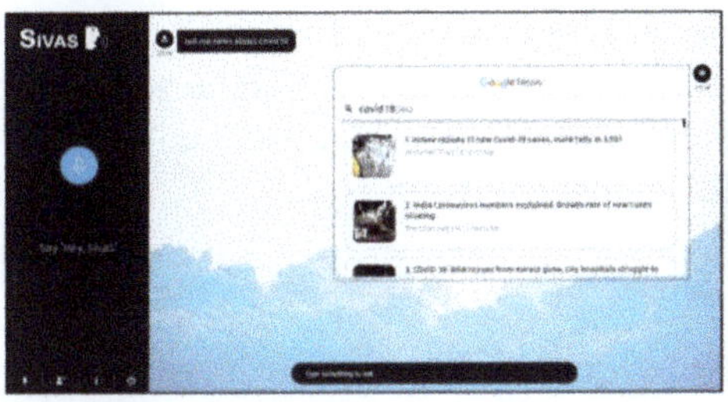

(e) Default - Show news for given topic

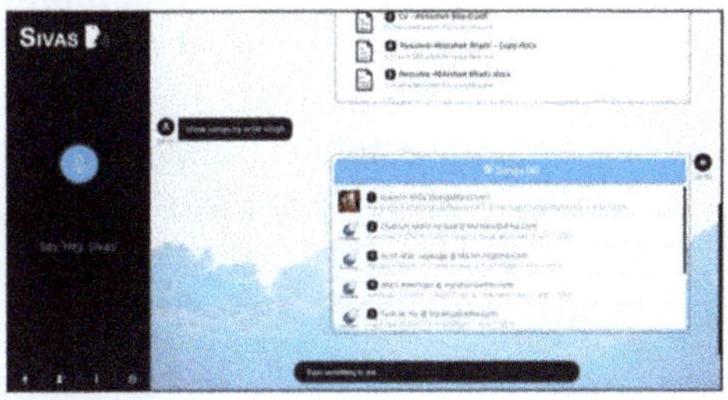

(f) Default - Show songs by artist name in local memory

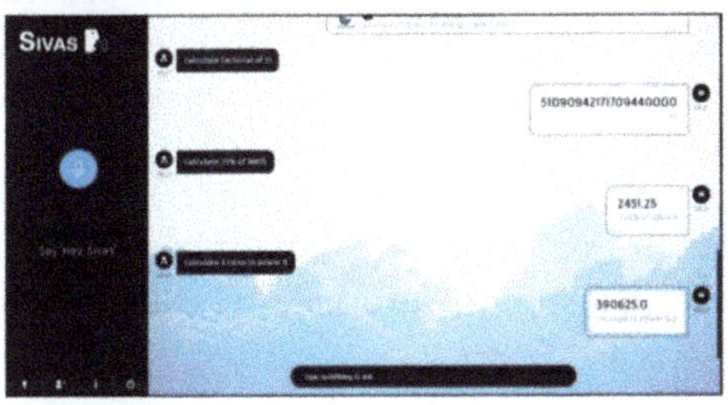

(g) Default - Solve some mathematical problems

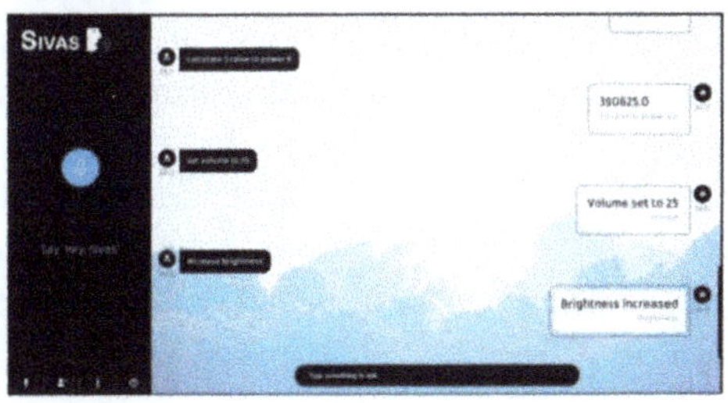

(h) Default - Change device volume and brightness

Figure 5.4. Outputs for the experimental queries listed above.

5.6 Experimental results

This section shows the outputs (see figure 5.4) obtained while experimenting and testing the application using various queries, listed below:

(a) Hey Sivas, ask dictionary to define quarantine

(b) Hey Sivas, ask Wikipedia to show details about New Delhi
(c) Hey Sivas, ask youtube to search videos of COVID 19
(d) Hey Sivas, search details of Sholay on IMDb
(e) Hey Sivas, tell me news about COVID 19
(f) Hey Sivas, show songs by Arijit Singh
(g) Hey Sivas, calculate factorial of 21
(h) Hey Sivas, calculate 25% of 9805
(i) Hey Sivas, calculate 5 raise to power 8
(j) Hey Sivas, set the volume to 25
(k) Hey Sivas, increase brightness

5.7 Limitations and future work

We would like to mention additional functionality that can be added to SIVAS in the future, so that its usability can be increased and it can be made more suitable for the next generation.

- Adaptive noise cancellation to suppress background noise in the user's utterance, in order to increase STT accuracy.
- Voice biometrics, which identifies the true owner of the device before processing commands to maintain the security and privacy of the user.
- Support can be extended to other natural languages and operating systems.
- Mood detection capability, so that the user's behavior can be detected and the system can actively contact the relevant authorities when needed.
- Addition of various third-party skills.

5.8 Result and conclusions

This chapter has presented a novel architecture for designing a smart and interactive VUI that can act as a virtual assistant and helps people to control their machines via voice while providing a hands-free, eyes-free experience. This architecture provides scaling flexibility, optimized techniques for using NLP in low-end devices, an efficient distribution of functionality, and effective ways of providing input. Our design resolves many drawbacks observed in current VUIs; it is therefore more reliable, usable, and scalable for the future. Regardless of all the above, helping people with disabilities in one way or another not only gives them happiness but also helps to uplift them in society; this is, therefore, a step toward helping these people.

Appendix A

Setup

In order to connect all the peripherals together and run the application, we need to configure the Raspberry Pi and place the required application code into its memory. Therefore, we need to perform a one-time configuration task, as follows:

(a) Connect all the hardware peripherals to their respective ports as shown in the block diagram.

(b) Start the power supply and the Raspberry Pi kit.
(c) Connect the Raspberry Pi to any other desktop or laptop via remote access.
(d) Once connected, open a terminal in Raspberry Pi’s Linux-based operating system,

Type the command *‘aplay -l,’* in order to check the speaker’s card and device number.
Type the command *‘arecord -l,’* in order to check the microphone’s card and device number.
Write the connection information for the speaker and microphone into a file *‘.asoundrc,’*

Type the command *‘vi .asoundrc’* and then press the key *‘I’* to start writing in the file.
Add the following code with your values for the speaker’s and microphone’s card and device number:

```
pcm.!default {
    type asym
    capture.pcm ‘mic’
    playback.pcm ‘speaker’
}
pcm.mic {
    type plug
    slave { pcm ‘hw:<card-number>, <device-number>’}
}
pcm.speaker {
    type plug
    slave {pcm ‘hw:<card-number>, <device-number>’}
}
```

Press the *‘ESC’* key and then type the command *‘:wq’* to save the updated file.

Install Python 3

Type the command ‘*sudo apt update*’
Type the command *‘sudo apt install python3’*

(e) Install the python application

Once all steps have completed, the configuration has been set up, and we can now run the application as follows:

(a) Run the python application
(b) Open the web browser, and access the URL ‘http://localhost:12000.’

Further reading

[1] Srinivas P, Sai Teja T, Bhavana C H, Likhith R and Sathish Kumar K 2020 Raspberry Pi based voice assistant using python *Int. J. Eng. Appl. Sci. Technol.* **4** 105–8 (https://ijisrt.com/assets/upload/submitted_files/1582975423.pdf)

[2] Datar P, Malewar P and Ramteke Y 2019 Voice enabled application for physically challenged individuals *Int. J. Res. Eng. Sci. Manag.* **2** 618–20
[3] Python Software Foundation 2020 *Python* https://python.org/
[4] MDN, JavaScript 2020 https://developer.mozilla.org/en-US/docs/Web/JavaScript
[5] MDN, HTML: Hypertext Markup Language 2020 https://developer.mozilla.org/en-US/docs/Web/HTML
[6] MDN, CSS: Cascading Style Sheets 2020 https://developer.mozilla.org/en-US/docs/Web/CSS
[7] Douglas Crockford, Introducing JSON 2020 https://json.org/json-en.html
[8] MDN, Speech Recognition 2020 https://developer.mozilla.org/en-US/docs/Web/API/SpeechRecognition
[9] Brainasoft, Braina 2020 https://brainasoft.com/braina/
[10] Amazon, Amazon 2020 www.amazon.com/alexa-skills/b?ie=UTF8&node=13727921011
[11] Gonfalonieri A 2020 *How Amazon Alexa works? Your guide to Natural Language Processing (AI)* https://towardsdatascience.com/how-amazon-alexa-works-your-guide-to-natural-language-processing-ai-7506004709d3
[12] PICOVOICE 2020 *Tips for Choosing a Wake Word* https://picovoice.ai/blog/tips-for-choosing-a-wake-word/
[13] Yordanov V 2020 *Introduction to Natural Language Processing for Text* https://towardsdatascience.com/introduction-to-natural-language-processing-for-text-df845750fb63
[14] Li S 2020 *Named Entity Recognition with NLTK and SpaCy* https://towardsdatascience.com/named-entity-recognition-with-nltk-and-spacy-8c4a7d88e7da
[15] Akshat J 2020 *A Brief Introduction to Intent Classification*
[16] Sweigart A 2020 *Automate the Boring Stuff with Python* https://automatetheboringstuff.com/
[17] Internet Engineering Task Force (IETF) 2020 *The WebSocket Protocol* https://tools.ietf.org/html/rfc6455
[18] MediaWiki 2020 *MediaWiki API* https://en.wikipedia.org/w/api.php
[19] Dictionary.com 2020 https://dictionary.com
[20] Thesaurus.com 2020 https://thesaurus.com
[21] Wikipedia 2020 https://en.wikipedia.org
[22] YouTube 2020 https://youtube.com
[23] IMDb 2020 https://imdb.com
[24] GoogleNews 2020 https://news.google.com
[25] Icons 2020 https://flaticons.net
[26] Raspberry Pi Foundation 2021 https://raspberrypi.org/
[27] Wikipedia 2021 Raspberry Pi https://en.wikipedia.org/wiki/Raspberry_Pi

IOP Publishing

Advances in Image and Data Processing using VLSI Design, Volume 1

Smart vision systems

Sandeep Saini, Kusum Lata, Abhishek Sharma and G R Sinha

Chapter 6

A topical survey of computing solutions for plant disease classification using deep learning techniques

Aniket Dhole, Mohit Gandhi, Shrishail Kumbhar, Harsh Singhal and Sonal Gore

Agriculture accounts for the majority of the production in the economy and also contributes to the gross domestic product (GDP). A major problem in agriculture is plant disease that is not recognized in the early stages, due to which, the people working in this industry face resulting losses, such as lost income, loss of time and effort, etc. Therefore, the authors have surveyed topical articles related to plant recognition and the identification of disease using computing solutions based on recent popular techniques, such as machine learning and deep learning. Along with the algorithms, we have also surveyed different hardware implementations of plant disease detection on embedded devices, such as the Raspberry Pi, field-programmable gate arrays (FPGAs) with very large scale integration (VLSI), and ARM processors that use frameworks such as TinyML and TFLite. The authors studied and analyzed major deep learning algorithms and techniques, such AlexNet, long short-term memory (LSTM), LeNet-5, and ResNet, which have been used for plant disease detection. Finally, the authors concluded a review to identify points of comparison and similarities. This chapter also summarizes the techniques used by existing methods along with their major challenges, which constitute research gaps in this particular area.

6.1 Introduction

Due to technological advancements, the ways in which we solve problems are changing day by day. Technologies such as machine learning, deep learning, artificial intelligence, and blockchain are growing daily and helping to solve many problems; agriculture is no exception to this process. Agriculture is an important

doi:10.1088/978-0-7503-3919-3ch6

aspect of the economy of any country. However, the crops or plants grown are being infected by many diseases, due to the use of artificial fertilizers and pesticides. The rate of production of diseased plants is growing at an alarming rate due to the lack of early disease detection. The agricultural population accounts for 67% [1] of the overall population, according to the Food and Agriculture Organization (FAO 2000 statistics). Agriculture accounts for 39.4% of the GDP and 43% of all exports. Almost 30% to 40% of the yield is unfortunately damaged each year due to various plant diseases. This, in turn, affects the overall economy. Therefore, there is a need for early detection of plant diseases.

The manual identification of plant diseases is very difficult, time-consuming, and tedious. In our advancing world, deep learning can help us to automate this technique and detect diseases rapidly without manual intervention. Modern advancements have helped researchers to detect diseases early with good results and accuracy. The use of deep learning (DL), machine learning (ML), and computer vision (CV) techniques can enable digital computers to identify any image data. With recent advancements in machine learning models using sensors and phone cameras, farmers can easily use these techniques on a small scale as well. Researchers have used these techniques to recognize diseases by scanning images of the fruits or leaves of a plant without human intrusion. Furthermore, these techniques can help to recommend steps to prevent these diseases by using the appropriate organic pesticides and fertilizers.

In this respect, our work provides a contrastive analysis of the different algorithms used for the identification of plant diseases, which could help farmers or researchers in the early stages of their careers.

6.2 Related work

Wallelign *et al* [2]. The authors focused on soybean plant disease, classifying diseases into four classes: healthy, *Septoria* blight, frogeye, and downy mildew. The DL algorithm used in this paper was LeNet. The 'PlantVillage' dataset was used, which provides soybean plant images of healthy and diseased plants. The dataset consisted of 12 673 samples, out of which, 6234 showed healthy leaves, 3565 illustrated *Septoria* blight, 2023 exhibited frogeye leaf spot, and 851 displayed downy mildew. All images were taken in uncontrolled environments, yet the system achieved a 99.32% accuracy. First, images of different resolutions were converted to 128 × 128 pixels. The proposed LeNet architecture was made up of three convolutional layers, max pooling was applied after each phase, and the final layer was fully connected. Each convolutional layer employed the rectified linear unit (ReLu) activation function. The first convolutional layer produced 32 kernels, each measuring 3 × 3 pixels. Following that, max pooling was implemented, and the result was used as the input of the second convolutional layer, which produced 64 kernels with a size of 4 × 4. A fully connected layer with 512 neurons followed the last convolutional layer, which had 128 kernels with a size of 1 × 1.

This algorithm was trained using adaptive moment estimation (Adam) with a batch size of 100 for 1000 epochs, and achieved an accuracy of 99.32%. The authors concluded that convolutional neural networks (CNNs) can extrapolate significant attributes from an uncontrolled environment, which is necessary for plant disease recognition.

Amara *et al* [3]. This paper illustrates the classification of banana plant diseases. The authors classified these into two diseases, namely black sigatoka and black speckle. The PlantVillage dataset was used for classification. Images were captured in an uncontrolled environment, resulting in 3700 samples. In the image preprocessing step, each image was converted to a size of 60 × 60 and then converted to gray scale. The training data were split in an 80%, 60%, 50%, 40%, 20% manner. Various accuracies were obtained, among which, the highest accuracy was 98.61%.

Turkoglu *et al* [4]. The authors created two models to classify apple diseases. One model was a single pretrained CNN model and the other model was a hybrid. The hybrid model used three pretrained CNN models fed to separate LSTM Layers. A softmax layer was then used to classify the score and the classes. The dataset used was collected from farms in Turkey; it contained four classes. The pretrained CNN models were GoogleNet, DenseNet201, and AlexNet. The accuracy achieved was 96.1% for individual models using DenseNet201 and 99.2% for hybrid models using all three pretrained CNNs. Each LSTM and softmax layer in the hybrid model had fully connected layers of AlexNet, GoogleNet, and DenseNet201, with a classification layer preceding the LSTM layer. The outputs of these three layers were computed using the majority voting classifier. This study also used automatic scheming to select the best parameters for training the model.

Sun *et al* [5]. Yield prediction is the biggest indicator in agriculture. Therefore, the authors proposed a DL model using remote sensing and CNN–LSTM to predict the yield of soybean plants at the country level. This is achieved using the Google Earth engine, weather, land surface temperature, previous yield, and a surface reflectance dataset. For the training, the authors processed the tensor layers of the Google Earth engine by adding data layers to the final image. They then trained them using their CNN–LSTM network. The root mean square error (RMSE) used for evaluation achieved a score of 329.53, which outperformed normal individual models based on CNN and LSTM.

Perugachi-Diaz *et al* [6]. In this study, convolutional neural networks were utilised to achieve the categorization of white cabbage seedling pictures. The authors focused on a dataset that tracks the growth of seedlings over 14 days and includes photos taken at four different points. The dataset was obtained from Bejo, an agricultural enterprise. It contains 13 200 unique plants with corresponding labels. The architecture used in this paper was AlexNet. AlexNet correctly classified 94% of the seedlings in the test set. The training, testing, and validation sets were partitioned using the ratio 3:1:1, which resulted in a split of 7920:2640:2640 examples from the available dataset.

Matin *et al* [7]. The authors focused on efficient disease detection in rice leaves using one of the DL neural networks. The DL algorithm used in this paper was AlexNet, which was used to detect three rice leaf diseases in three classes known as leaf smut, brown spot, and bacterial blight. The dataset used was Rice Leaf

Disease Detection, downloaded from kaggle.com, which contains three files of rice leaf disease with 40 images in each file.

A very small dataset of 120 images was further increased to 900 images using image augmentation. Seventy percent of the data (630 images) were used as training images and 30% of the data (270 images) were used as test images. The RGB input image size was $227 \times 227 \times 3$, which was transmitted to the first convolutional layer, which filtered the picture using 96 kernels of size $11 \times 11 \times 3$ with a stride of 4 pixels, followed by the ReLu activation function after each convolutional layer. Max pooling was applied only after the first two convolutional layers, whose outputs were respectively passed onto the following layers. The final output was generated using softmax and a classification layer. An accuracy of 99.42% was obtained by the authors using AlexNet for rice leaf disease detection. The authors concluded that AlexNet had improved accuracy compared to the other approaches used for the detection of rice leaf diseases and that in the future, the K-means clustering algorithm, fuzzy c-mean clustering could be used, because of its accuracy and processing time.

Rangarajan *et al* [8]. Early disease detection utilising a rapid, reliable, non-destructive technology would aid farmers to combat the problem of disease. The authors employed pretrained DL algorithms to classify tomato crop illnesses. Images of tomato leaves were classified into six diseased classes: leaf mold, late blight, target spot, two-spotted spider mite attack, tomato yellow leaf curl virus, and tomato mosaic virus and a healthy class from the PlantVillage dataset. These were fed into AlexNet and VGG16 net, two DL architectures. For 13 262 images, the recognition rates using VGG16 and AlexNet were 97.23% and 97.49%, respectively.

Qi *et al* [9]. The authors focused on peanut leaf disease. Peanuts are an important crop extensively used in day-to-day life. Peanut leaf disease can hamper the growth and production of peanuts. On researching peanut leaf disease, the authors found 3–4 disease patterns, consisting of leaf rust, leaf scorch, combined rust and scorch on leaves, and leaf spot. The authors used both traditional machine learning and deep learning models to detect diseased peanut leaves. The dataset used was the open source PlantVillage dataset, which includes all the above leaf patterns. After exploring, analyzing, and cleansing the data, the authors had 6029 peanut leaf patterns in total. The models used included five CNN models: AlexNet, ResNet50, DenseNet121, VGG16, Inception v1, and two traditional ML models: support vector machine (SVM) and logistic regression.

Among all the models, the CNN models (deep learning models) had considerably better accuracy than the traditional ML models. The model accuracy details are as follows: DenseNet121: 95.97%, ResNet50: 95.25%, Inception V1: 95.07%, VGG16: 93.35%, AlexNet: 91.79%, SVM: 73.20%, and logistic regression: 61.33%. In conclusion, the authors found that the deep learning models outperformed typical machine learning models. ResNet50 and DenseNet121 both had the highest accuracy.

Xiao *et al* [10]. The paper discusses strawberry leaf disease identification. The authors employed three distinct CNN models: Renet50, VGG16, and GoogleNet. Blight on the leaves (rotting of both crown and fruit), gray fungus, gritty mildew, and a combination of the two diseases were discovered. ResNet50 had a 100% accuracy rate for blight on leaves (rotting crown) and a 98% accuracy rate for gray fungus and grainy mildew disease. GoogleNet had a 100% classification rate for grainy mildew, while VGG16 had a 100% classification rate for gray fungus disease. For all three models discussed above, the accuracy of detection of blight on the leaves for rotten fruits was 100%. The authors concluded that different CNN models have different classification rates for various disease patterns found in strawberry leaves.

Ingale ***et al*** [11]. The authors used a new plant disease dataset, which is freely available on the Kaggle website and contains approximately 80 000 or more color images. The objective was to distinguish between normal plant leaves and diseased plant leaves on a variety of plants. It took the authors about 14 plants to accomplish this. Approximately 26 diseases were discovered during the process; each plant had a unique disease.

Liu ***et al*** [12]. This paper contains a survey of plant disease recognition and pest detection. The authors surveyed the work done on this topic and listed challenges such as small datasets, data amplification, and in particular, how the data are gathered. The work done on this topic was not universal, as not all plant diseases are covered yet, so there is a lot of scope to work on this topic.

Abade ***et al*** [13]. The authors reviewed 121 research papers and analyzed them using many parameters, such as dataset, approach, algorithm, etc. This study provides a collection of all approaches in one place.

Maniyath ***et al*** [14]. The authors focused on disease detection using machine learning using large datasets and custom datasets. Image preprocessing is the process of reducing all images to the same size. The characteristics are then extracted from the preprocessed images, which is accomplished with the assistance of Histogram of Oriented Gradient. The objective of the random forest classifier proposed in the paper was to identify and detect disease abnormalities which are found on plants in the natural environment. An accuracy of 70% was achieved using the proposed model. It is possible to improve the accuracy by using a vast number of samples and adding more features.

Sharma ***et al*** [15]. The authors experimented with various CNN models and datasets to check the consistency and accuracy of these models on each algorithm with the same dataset. The AlexNet, GoogleNet, and ResNet50 models were used. The datasets used were CIFAR10 and CIFAR100 with ten classes each. For the CIFAR100 dataset, the accuracies were: AlexNet—44.10%, GoogleNet—64.40%, and ResNet50—59.82%. For the CIFAR10 dataset, the accuracies were: AlexNet—36.12%, GoogleNet—71.67%, and ResNet50—78.10%.

Akila ***et al*** [16]. The authors applied three methods to detect plant illness using leaf images: a faster region-based convolutional neural network (Faster R-CNN), a region-based fully convolutional network (R-FCN), and a single-shot multibox detector (SSD). Many crops, such as rice, wheat, cotton, potato, sugarcane, chili, carrot, banana, guava, and brinjal (eggplant) were included in the dataset, which was gathered from the internet and photographs recorded using cameras. All images were manually annotated with a position of the disease and class of the disease along with the plant name. For training, all images were resized to 256 × 256 for faster training. Faster R-CNN was used for object recognition along with the region to predict the class and location contained in the image. R-FCN uses a region proposed network to create an image and crop features from the last layer. The SSD selects the top feature and high-resolution feature map at lower resolution. For feature extraction, the parameters were calculated using average precision (AP) and the intersection-over-union (IoU). The authors concluded that their model successfully recognized various plant diseases and gave solutions for those diseases.

Rajesh Kumar *et al* [23]. Here, the authors created a binary convolutional neural network (BCNN) using a field-programmable gate array (FPGA) architecture based on a Spartan-3 Xilinx FPGA board. The proposed architecture consisted of binary weight kernel convolution and neuron scaling along with ReLU and batch normalisation. They used the BCNN to control the rate of a DC motor using sensor data.

Syamsuri *et al* [24]. In this paper, the authors used the PlantVillage dataset to classify coffee plant disease. The authors also used mobile devices to take pictures and check for disease. The authors used MobileNet and Mobile Nasnet (MNasNet). MNasNet achieved an accuracy of 94.87%, while MobileNet achieved an accuracy of 92.83%. The time latency for MobileNet was low, at 394.70 ms, and that of MNasNet was 430.20 ms.

Emebo *et al* [25]. The authors used a deep convolutional neural network (DCNN) as a classification technique with the Keras and Tensorflow Python machine learning libraries to develop a model for on-site plant disease detection using a hand-held Raspberry Pi device. The authors' major focus was to use a CNN to detect tomato *Septoria* leaf spot and tomato mosaic infections. The PlantVillage dataset was used in this study. With a validation accuracy of 99.01%, the CNN was able to automatically extract relevant characteristics from photos.

Table 6.1 provides an overview of the various papers, including parameters such as the dataset used and the performance.

Table 6.2 compares the articles in terms of parameters such as the number of classes, training hyperparameters, preprocessing techniques, experimental details, etc.

6.3 Algorithmic survey

Models that roughly correspond to neurons in the brain are known as neural networks. Every neuron in a neural network takes input, performs operations, and outputs results that are passed on as the input to the next neuron. The study of computers running these neural networks for various applications is known as DL; in this approach, a machine passes data to the network and receives output in the form of numbers or a probability. In image classification, the image is first converted into an array of pixel data and passed to the first layer of the neural network. Using algorithms such as CNN, Alexnet, LeNet, Resnet, etc. (the further architecture varies) data is then passed between layers, updating the weights and biases of each node in the network and producing an output, either as classes or as a probability, using activation functions such as softmax, ReLu, etc.

Table 6.1. Comparative study of research articles based on model, dataset, and performance (accuracy).

Authors	Objective(s)	Deep learning model(s) used	Dataset	Accuracy
Wallelign *et al* [2]	To detect soybean plant disease	LeNet-5	PlantVillage	99.32%
Amara *et al* [3]	Classification of banana plant disease	LeNet-5	PlantVillage	98.61%
Turkoglu *et al* [4]	To classify apple disease using two different models	CNN–LSTM	Real-time data from Turkish farms	Hybrid: 99.2%, individual: 96.1%
Sun *et al* [5]	Soybean yield prediction	CNN–LSTM	MODIS SR data, MODIS LST data	RMSE value: 329.5
Perugachi-Diaz *et al* [6]	Classification of white cabbage seedlings	AlexNet	Bejo Organisation	94%
Matin *et al* [7]	Detection of diseases of rice leaves	AlexNet	Rice Leaf Disease Detection	99.42%
Rangarajan *et al* [8]	Classification of tomato disease using deep learning algorithms	AlexNet, VGG16	PlantVillage	Alexnet: 97.49%, VGG16%: 97.23%
Qi *et al* [9]	Peanut leaf disease identification based on stack ensemble	ResNet	Manually collected at South China Agricultural University	ResNet50%: 97.59%, DenseNet121%: 90.50%
Xiao *et al* [10]	Detection of strawberry diseases	ResNet	Strawberry farm at Dahu township	Original dataset: 98.06%, with features: 99.60%

Table 6.2. Comparative study of research articles based on classes, preprocessing techniques, model hyperparameters, and experimental details.

Authors	Output classes	Preprocessing technique	Deep learning model hyperparameters	Experimental details
Wallelign [2]	Healthy, *Septoria* blight, frogeye, and downy mildew	All images are resized to 128 × 128	Three convolutional layers, max pooling, ReLu, number of kernels = 32, 64, 128. Batch size = 100. Number of epochs = 1000. Adam optimizer	70% training, 10% validation, 20% testing
Amara *et al* [3]	Healthy, black sigatoka, black speckle	All images are resized to 60 × 60 and converted to gray scale	Three convolutional layers, max pooling, ReLu	80% training and 20% testing; 60% training and 40% testing; 50% training and 50% testing; 40% training and 60% testing; 20% training and 80% testing
Turkoglu *et al* [4]	*Venturia inaequalis*, *Monilinia laxa*, *Eriosoma lanigerum*, *Aphis pomi*		Batch size = 10 100, epochs = 10 100, number of neurons in LSTM = 100, 1500. RMSProp, Adam optimizer	75% training and 25% testing
Sun *et al* [5]		Image layers are masked	Batch size = 16, epochs = 100, number of neurons in LSTM = 256. ReLU optimizer	

Perugachi-Diaz *et al* [6]		Resized to a resolution of 64 × 64	Five convolutional layers. Batch size = 128, epochs = 500. Adam, stochastic gradient descent (SGD) optimizer	60% training, 20% validation, 20% testing
Matin *et al* [7]	Bacterial blight, brown spot, and leaf smut	Image augmentation is performed to increase sample size	Three convolutional layers, max pooling, ReLu, number of kernels = 96	70% training and 30% testing
Rangarajan *et al* [8]	Late blight, leaf mold, two-spotted spider mite attack, target spot, tomato mosaic virus disease, tomato yellow leaf curl virus disease, and healthy	Input image is augmented to 227 × 227 pixels	Five convolution layers, batch size = 32, epochs = 10, ReLu activation	
Qi *et al* [9]	Scorch, rust, a combination of scorch and rust, leaf spot	All types of image converted to RGB format and a size of 224 × 224 pixels. Data augmentation is performed.	50 convolution layers; kernels used = 64, 128, 256, 512, 1024, 2048; epochs = 20. ReLu activation	80% training and 20% testing
Xiao *et al* [10]	Blight on leaves (crown and fruit rotten), gray fungus, and gritty mildew	Data augmentation is performed.	50 convolution layers; kernels used: 64, 128, 256, 512, 1024, 2048. ReLu activation	80% training and 20% testing

6.3.1 LeNet-5 [17, 18]

LeNet-5 was known as the first application in the image processing field and was developed in 1998 by Yann LeCun. Its objective was to classify gray scale images of digits taken from the Modified National Institute of Standards and Technology Database (MNIST), which consists of images with dimensions of 32 × 32 × 1. The LeNet-5 architecture consists of seven layers (or eight including the input layer); the seven layers consist of three convolutional layers, two subsampling layers, and two fully connected layers. A LeNet architecture is shown in figure 6.1.

The architecture contains two main layers: one is the convolutional layer and the other is the subsampling layer. To reduce the computational complexity, the dimensions of images were reduced in the subsampling layer so that the model would depend more on features than size. A description of all the layers is as follows:

The convolutional layer is denoted by C_x, the subsampling layer by S_x, and the fully connected layer is denoted by F_x, where x represents the corresponding position of that layer in the architecture.

- Input layer: since noting is studied in this layer, it is not considered in the architecture. This layer takes a 32 × 32 image as an input; if the image has a different size, we have to preprocess that image to add padding around the image to make it 32 × 32 in size.
- The output of C1 is six feature maps, each 28 × 28 in size, with a kernel of size 5 × 5; C1 has 156 trainable parameters and 122 304 connections.
- S2 has six feature maps, each 14 × 14 in size, i.e. half the size of C1's feature map. It has 5880 connections and 12 trainable variables.
- C3 generates 16 feature maps; each has a kernel size of 5 × 5. C3 has 151 600 connections and 1516 trainable variables.
- LeNet-5 [17, 18] S4 includes 16 feature maps, each with a kernel size of 5 × 5. It has 2000 connections and 32 trainable properties.
- C5 is the last convolutional layer; it has 120 feature maps. As the size of S4 is 5 × 5, C5 is connected to all 16 feature maps of S4. Therefore, the C5 feature map is 1 × 1 in size. It has 48 120 trainable connections.
- F6 is made up of 84 units, or neurons. It features 10 164 trainable parameters and is fully coupled to C5.

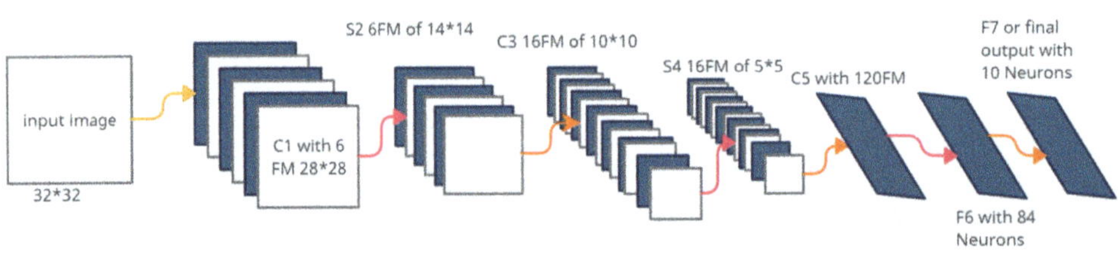

Figure 6.1. LeNet architecture.

- F7, or the final output layer, contains ten neurons, corresponding to the number of classes present in the MNIST dataset that represent digits (0–9).

Here, 'kernel' is a name given to a window or a matrix. The kernel contains the values of the weights which are used to convolve the input image. The main function of the subsampling layer is to reduce the feature map to halves which ensures good computational complexity. This process is also known as downsampling. Two more layers that are used in networks are the flattening layer and the dense layer. The purpose of the flattening layer is to convert the input into a one-dimensional array; the dense layer contains the hidden units or neurons. The activation function used to activate neurons is the softmax activation function or tanh. Softmax activation is the next or generalized version of the sigmoid activation function. In conclusion, LeNet was the first network to be used for image recognition; it is simple to understand and has a fast training time. Its major advantages are that it is simple to learn and fast.

6.3.2 Convolutional neural network–long short-term network (CNN–LSTM) [19]

CNNs (convolutional neural networks) are used in modeling problems related to spatial inputs, such as images. Computer vision, picture classification, and object detection have all proved successful using CNNs. Long short-term memories (LSTMs) are utilised in modelling sequences and predictions based on them. LSTMs are commonly used in NLP-related tasks, such as machine translation, sentence categorization, and generation. A CNN–LSTM is the integration of a CNN with an LSTM; it was developed in 2015. To learn the temporal properties, the output of the CNN output is multiplied by the contents of the LSTM cell.

The CNN–LSTM architecture extracts features from input data using CNN layers, which are passed to LSTMs, thereby supporting sequence prediction. It was created for sequence prediction problems with spatial inputs, such as photos or videos. The model's architecture is displayed in figure 6.2.

The details of the model's architecture are as follows:

- The initial layer uses two CNN layers, generally with 256 feature maps.
- A non-overlapping max-pooling strategy is used; it has a size of three for the first layer and zero for the second CNN layer.
- A linear layer is then used to reduce the output size (to 256 layers) of the CNN layer.

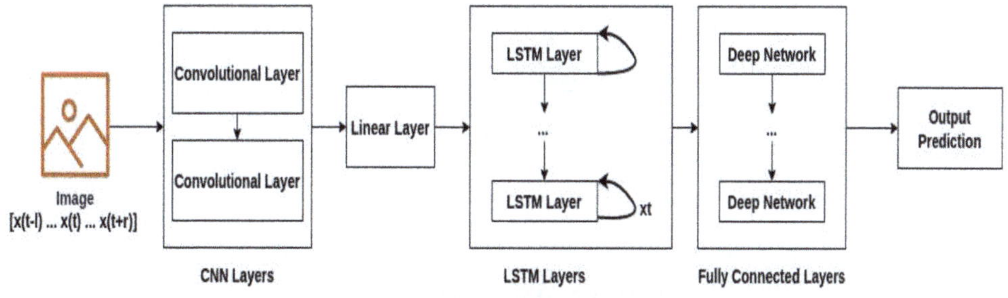

Figure 6.2. CNN–LSTM architecture.

- Uses two LSTM layers with 832 cells and a 512-unit projection layer to reduce the size.
- The LSTMs then pass the series data for some frequency frames.
- Finally, we pass the data to deep network layers for the generation of feature-rich output which is easy to recognize and separate.

The advantages of this model are:

- CNNs are used to extract features from audio and textual input data for LSTMs in this architecture, which has been used for speech recognition and natural language processing problems.
- It provides fast results for serial images, such as video, as the LSTMs store important features from previous data.

The disadvantages of this model are:

- Overfitting is a problem with LSTMs, and using the dropout algorithm to solve it is difficult. While training a network, dropout is a regularization method in which input and recurrent connections to LSTM units are probabilistically excluded from activation and weight updates.
- Different random weight initializations affect LSTMs, making them behave similarly to a feed-forward neural network. Lightweight initializations are preferred instead.

6.3.3 AlexNet [20, 21]

AlexNet is a convolutional neural network that has had a significant impact on machine learning, particularly in the field of deep learning for machine vision. AlexNet was the winning entry in the ImageNet Large Scale Visual Recognition Challenge 2012 (ILSVRC2012). AlexNet solved the image classification

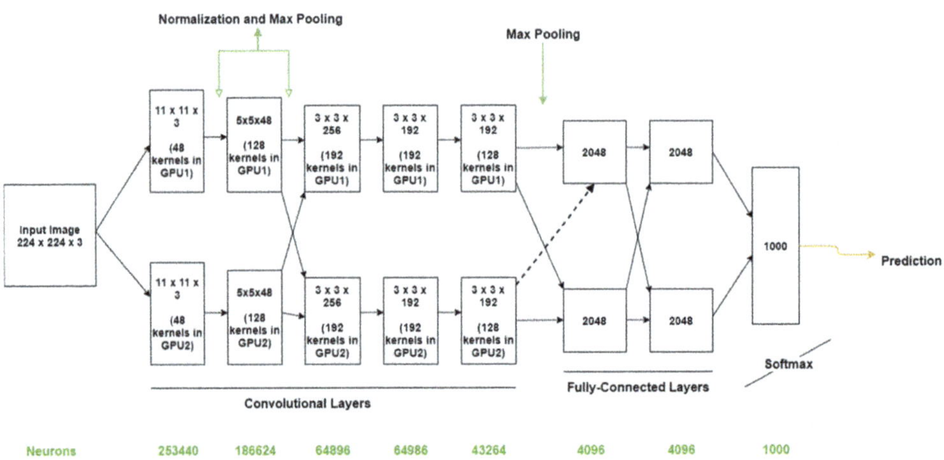

Figure 6.3. AlexNet architecture.

challenge by taking one of 1000 possible classes as its input and producing a vector of 1000 integers as its output. AlexNet uses the ImageNet dataset, which has 15 million labeled high-resolution images. The model architecture is shown in figure 6.3.

AlexNet has eight layers, starting with five convolutional layers and ending with three fully linked layers. The input to AlexNet is an RGB image 256 × 256 in size. If the image size is not 256 × 256, it must be converted before using AlexNet.

- If the source image is gray scale, it is transformed to RGB by reproducing the single channel to create a three-channel RGB image.
- To obtain the input for the first convolutional layer, a 227 × 227 arbitrary crop is made from the 256 × 256 image.
- Using 96 kernels with a size of 11 113 and a stride of 4 pixels, the first convolutional layer filters the 2 242 243 input images. (Two GPUs are used; each trains 48 kernels.)
- The output of the first convolutional layer is fed into the second convolutional layer, which is normalised, pooled, and then filtered using 256 kernels that are 5 × 5 × 48 in size.
- There are no pooling or normalising layers between the third, fourth, and fifth convolutional layers.
- 384 kernels with a size of 3 × 3 × 256 are connected (normalised and pooled) to the outputs of the second convolutional layer in the third convolutional layer.
- The fourth convolutional layer contains 384 3 × 3 × 192 kernels, while the fifth convolutional layer contains 256 × 3 × 3 192 kernels.
- Each of the completely connected layers contains 4096 neurons.
- The last layer's output is sent into a 1000-way softmax, which generates a distribution over the 1000 class labels.

The top-1 performance of the network suffers a loss of around 2% when any of the intermediary levels are removed. After the convolutional and completely connected layers, ReLu nonlinearity is used. Nonlinear data can be taught faster using ReLu than by activation functions, such as tanh or sigmoid functions. Pooling layers are commonly used to downsample the tensors' widths and heights while maintaining the same depth. $f(x) = \tanh(x)$ or $f(x) = (1 + e - x) - 1$ are common formulas for modelling a neuron's output f as a function of its input x.

These saturating nonlinearities are substantially slower than the non-saturating nonlinearity $f(x) = \max$, in terms of training time with gradient descent $(0, x)$.

The advantages of this model are:

- AlexNet was the first large CNN model trained using GPUs. As a result, models may be trained faster.
- When compared to LeNet, AlexNet has a deeper architecture with eight layers, which means it is better at extracting features. It also works nicely with colour photos at the moment.
- AlexNet's ReLu activation function provides some advantages. Unlike other activation functions, it does not limit the output. As a result, no features are lost.

It can be concluded that AlexNet is much better than the LeNet algorithm in terms of efficiency and many other features.

6.3.4 ResNet [22]

ResNet is a well-known model for handling exploding or vanishing gradient issues. ResNet involves the use of a short-cut connection concept. The output obtained from preceding layers is successfully added to the output of successive layers in the residual block, therefore making ResNet much more accurate than other models. The model's architecture is shown in figure 6.4.

- ResNet50, as its name says, is ResNet with a depth of 50 layers.
- The 50 layers of the ResNet50 model are a combination of convolution layers, a max-pool layer, an average pooling layer, and a fully connected layer.
- First, the convolution used in the ResNet50 architecture has a 7 × 7 kernel size; about 64 kernels form the first layer.
- Following convolution, the next layer is the max-pool layer with a size of two.
- Following this are convolutions of 1 × 1 and 3 × 3, both with 64 kernels and a 1 × 1 convolution with 256 kernels. These three layers are repeated three times to form up to nine layers in ResNet50.
- Following this are convolutions of 1 × 1 and 3 × 3, both with 128 kernels and a 1 × 1 convolution with 512 kernels; these three layers are repeated of four times to form up to 12 layers in ResNet50.
- Following this are convolutions of 1 × 1 and 3 × 3, both with 256 kernels and a 1 × 1 convolution with 1024 kernels; these three layers are repeated six times to form up to 18 layers in ResNet50.
- Following this are convolutions of 1 × 1 and 3 × 3, both with 512 kernels and a 1 × 1 convolution with 2048 kernels; these three layers are repeated three times to form up to nine layers in ResNet50.
- Finally, a completely linked layer consisting of 1000 types of object forms up to one layer followed the average pool layer (the fiftieth layer).
- ResNet with 50 layers skips about three layers, while ResNets with 18 and 34 layers skip about two layers during the working stage.

The advantages of this model are:

- The ResNet model helps to deal with the vanishing gradient problem, as its error rate tends to decrease with increasing network depth; the reverse is the case for a plain network.

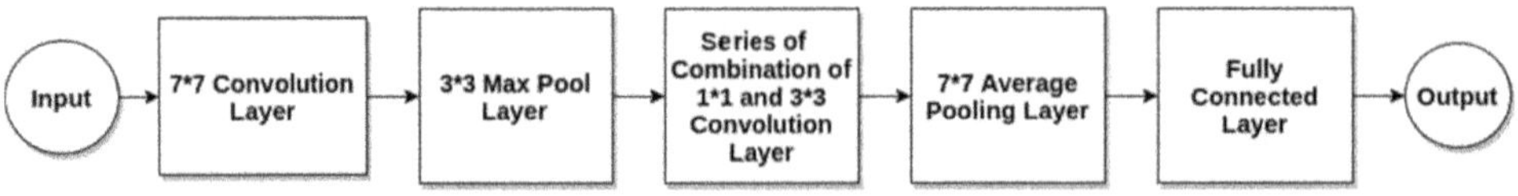

Figure 6.4. ResNet50 architecture.

- ResNet makes use of identity mapping or simple short-cut connection concepts, which increases the accuracy and efficiency of the model as compared to others.

The disadvantages of this model are:

- ResNet is suitable only for a depth of about 150 layers. Performance gets worse when the number of layers exceeds 150–200 layers.
- ResNet involves the use of a batch normalization layer which is, indeed, a complex task, and ResNet depends on it heavily.

6.4 Comparative study for performance analysis

The bar graph shown in figure 6.5 compares the performances (accuracies) of the different models described in the various papers.

The bar graph shows how the accuracy of the algorithm varies across the papers. The first two bars display the accuracy of the LeNet-5 architectures used in papers [2] and [3] for soybean and banana plant disease detection, respectively (both used the PlantVillage dataset). The soybean study achieved greater accuracy than that of banana plant diseases. The next two bars display the accuracy of CNN–LSTM architecture used in [4]; the authors used two models, one of which was a hybrid and the other of which was an individual model; among these, the hybrid model exhibited more accuracy. The next three plots show the accuracy of AlexNet, as used in three papers [6–8]. The comparison (figure 6.6) shows that [7] achieved the highest accuracy. The next two plots display the accuracy of ResNet50, which was used in papers [9] and [10]. The model described in [10] has the highest accuracy of all the algorithms.

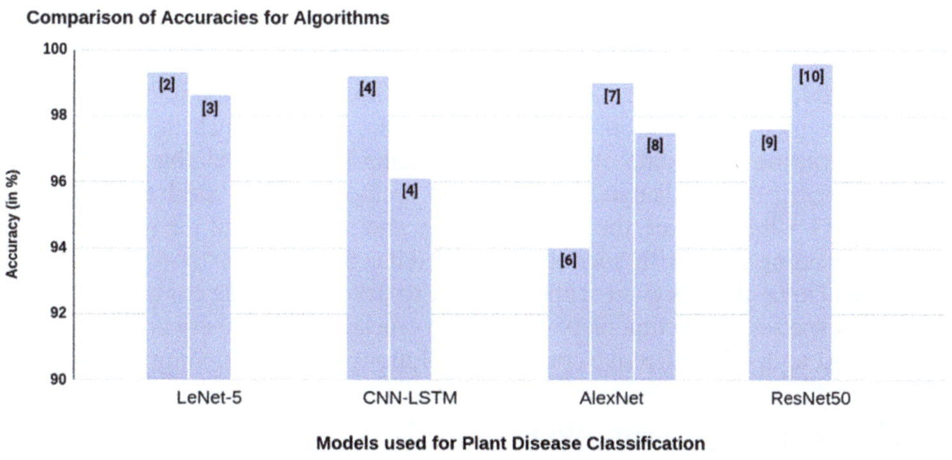

Figure 6.5. Comparison of the performance (accuracy) of different models described by different papers. The research papers are cited and shown at the top of each bar.

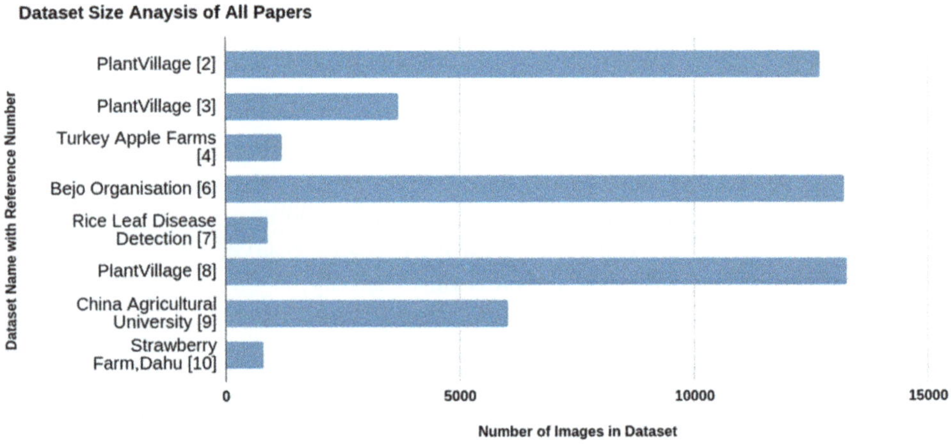

Figure 6.6. A comparative plot of the dataset sizes (number of samples) used in the various articles.

6.5 Common findings and challenges

6.5.1 Common findings

All the above algorithms are categorized as different types of CNN. CNN can be used for both data preprocessing and feature extraction. Today, CNN is commonly used for detection-related research, so the popularity of CNN algorithms is rapidly growing. Almost all papers that used real-time datasets demonstrated greater accuracy than the techniques used in those papers in which datasets were collected under uniform conditions. Almost all the techniques in most of the articles reduced the size of the dataset by resizing images for faster processing. In the detection of plant diseases, all papers have used hand-crafted feature extraction to extract disease spots using CNN layers.

6.5.2 Challenges

The main challenge for any computer vision task is to obtain a precise dataset for each type of plant disease. Many of these papers have collected their data in real time, while others have used a common, publicly available dataset, such as PlantVillage. Experiments using small datasets find it challenging to obtain reliable and validated results. To solve this challenge, augmentation methods can be used to increase the dataset size [7–10]. However, the generalizability and robustness of the model becomes another cascading issue with such augmented data. Since species may vary across a geographical area, there is great scope for work to develop a single computing solution for all plant species. Predicting plant disease is a crucial task and there is much scope to innovate and to develop effective computing solutions to help the community at large.

6.6 Hardware implementations

Based on the papers studied, the various hardware devices that can be used for plant disease detection are:

1. FPGA—there are currently many FPGA boards, such as the Xilinx Spartan, Zynq, Atrix, and Intel Cyclone that support machine vision. To run computer vision models on an FPGA, we need to design the model using binary weights for reduced memory use, instead of floating-point values. Since FPGAs are already networks of gates, binary weights can easily be applied to produce an output. Many new and complex methods have been prebuilt for newer FPGAs to increase accuracy. The other advantages of using FPGAs are reduced latency time when switching processes, compared to CPUs; this speeds up the process by 20 times.
2. Raspberry Pi—using the Raspberry Pi, we can create a complete portable system to support computer vision in the detection of plant disease, as we can attach cameras and sensors directly to the device and use the easy graphical user interface of the operating system with Python and a powerful ARM-based architecture. Here, any pretrained or custom CNN model can be used to process images from a camera and we can get decent frame rate of around 10–15 frames per second. We can also attach computing modules, such as Google Coral and Intel Neural Stick, to increase the performance of the model.
3. Android—Tensorflow is essential for conversion in order to employ mobile devices in detection operations. The model is first trained on a PC, after which, the .pb file is converted to a .lite file and copied to the mobile device. The operating system for mobile devices should be Android 8.0 and the device should have 4GB of RAM. A camera with at least 12 megapixels is required to capture a disease image. The image is fed into the TFlite model, which generates the output.

6.7 Conclusions

DL is very useful, and it can help experts such as biologists and ecologists in their work of studying and conserving the environment. Different algorithms and pretrained deep models, such as AlexNet, LeNet, ResNet50, etc. can be used for fast and efficient plant disease detection at an early stage. Using such algorithms, we can also explore the causes of such diseases to completely prevent them in the future. With enhancements in embedded AI, we can port such deep models to small devices and take advantage of those everywhere at a low cost. Following this research analysis, we can propose the development of a hybrid deep learning model with an appropriate selection of various hyperparameters in order to achieve better performance than those of existing models.

Bibliography

[1] The Food and Agriculture Organization (FAO) 2019 http://fao.org/news/story/en/item/1187738

[2] Wallelign S, Polceanu M and Buche C 2018 Soybean Plant Disease Identification Using Convolutional Neural Network *31st Int. Florida Artificial Intelligence Research Society Conf.* (Palo Alto, CA: AAAI Press) p 146

[3] Amara J, Bouaziz B and Algergawy A March 2017 A Deep Learning-based Approach for Banana Leaf Diseases Classification *Datenbanksysteme für Business, Technologie und Web (BTW 2017) (Lecture Notes in Informatics (LNI), Gesellschaft für Informatik)* (Bonn: Gesellschaft für Informatik) p 79
[4] Turkoglu M, Hanbay D and Sengur A 2019 *Multi-Model LSTM-Based Convolutional Neural Networks for Detection of Apple Diseases and Pests* (Berlin: Springer)
[5] Sun J, Di L, Sun Z, Shen Y and Lai Z 2019 County-level soybean yield prediction using deep CNN-LSTM model *Sensors* **19** 4363
[6] Perugachi-Diaz Y, Tomczak J M and Bhulai S May 2021 Deep learning for white cabbage seedling prediction *Comp. Elec. Agri.* **184** 106059
[7] Matin M H Md., Khatun A, Moazzam G Md. and Uddin M S 2021 An efficient disease detection technique of rice leaf using AlexNet *J. Comp. Commun.* **8** 49–57
[8] Rangarajan A K, Purushothaman R and Ramesh A 2018 Tomato crop disease classification using pre-trained deep learning algorithm *Proc. Comp. Sci.* **133** 1040–47
[9] Qi H, Liang Y, Ding Q C and Zou J 2021 Automatic identification of peanut-leaf diseases based on stack ensemble *Appl. Sci.* **11** 1950
[10] Xiao J-R, Chung P-C, Wu H-Y, Phan Q-H, Andrew Yeh J-L and Hou M T-K 2020 Detection of strawberry diseases using a convolutional neural network *Plants* **10** 31
[11] Ingale A 2020 *Plant disease detection using ResNet* Medium
[12] Liu J and Wang X 2021 Plant diseases and pests detection based on deep learning: a review *Plant Methods* **17** 22
[13] Abade A S, Ferreira P A and de Barros Vidal F 2021 Plant diseases recognition on images using convolutional neural networks: a systematic review *Comput. Electron. Agri* **185** 106125
[14] Ramesh Maniyath S, Vinod P V, Niveditha M and Pooja R *et al* 2018 Plant disease detection Using machine learning *Int. Conf. on Design Innovations for 3Cs Compute Communicate Control (ICDI3C)* pp 41–5
[15] Sharma N, Jain V and Mishra A 2018 An analysis of convolutional neural networks for image classification *Proc. Comp. Sci* **132** 377–84
[16] Akila M and Deepan P 2018 Detection and classification of plant leaf diseases by using deep learning algorithm *Int. J. Eng. Res. Technol. (IJERT)* **6** IJERTCONV6IS07082
[17] Rosebrock A 2016 *LeNet—Convolutional Neural Network in Python* https://pyimagesearch.com/2016/08/01/lenet-convolutional-neural-network-in-python/
[18] Agarwal S 2020 *LeNet-5 CNN Architecture* https://medium.com/@coolshaleen1/lenet-5-cnn-architecture-468ed7b895cd
[19] Kang E 2017 *Long Short-Term Memory (LSTM): Concept* https://medium.com/@kangeugine/long-short-term-memory-lstm-concept-cb3283934359
[20] Nayak S 2018 *Understanding AlexNet* https://learnopencv.com/understanding-alexnet
[21] Wei J 2019 https://towardsdatascience.com/alexnet-the-architecture-that-challenged-cnns-e406d5297951
[22] Dwivedi P 2019 *Understanding and Coding a ResNet in Keras* https://towardsdatascience.com/understanding-and-coding-a-resnet-in-keras-446d7ff84d33
[23] Rajesh Kumar C J, Vinod Kumar D, Baskar D, Mary Arunsi B, Jenova R and Majid M A 2019 VLSI design and implementation of high-performance binary-weighted convolutional artificial neural networks for embedded vision based Internet of Things (IoT) *Proc. Comp. Sci.* **163** 639–47

[24] Syamsuri B and Kusuma G P 2019 Plant disease classification using lite pretrained deep convolutional neural network on android mobile device *Int. J. Innov. Technol. Explor. Eng. (IJITEE)* **9** 2796–804

[25] Emebo O, Fori B, Victor G and Zannu T 2019 Development of tomato septoria leaf spot and tomato mosaic diseases detection device using raspberry Pi and deep convolutional neural networks *J. Phys.: Conf. Ser.* **1299** 012118

IOP Publishing

Advances in Image and Data Processing using VLSI Design, Volume 1

Smart vision systems

Sandeep Saini, Kusum Lata, Abhishek Sharma and G R Sinha

Chapter 7

Hardware IP cores for image processing functions

Anirban Sengupta and Rahul Chaurasia

Computationally intensive functions, such as image processing, video processing, etc., as used in consumer electronics systems, require huge amounts of data processing and incur significant power consumption. Therefore, dedicated hardware intellectual property (IP) cores can play a pivotal role for such data-intensive applications. This chapter discusses the high-level synthesis (HLS) design methodology as applied to the dedicated optimized hardware IP cores used for five image processing functions: (a) the Gaussian blur filter, (b) the Sobel edge detector, (c) the Prewitt gradient edge detector, (d) image magnification, and (e) image contrast/brightness adjustment. Case studies of the methodology show that the designed hardware IP cores have smaller gate counts and require less power than their non-optimized counterparts.

7.1 Introduction

Image processing functions find wide utility in many applications, such as robotic vision, biometric security, medical imagery, and so forth. The aim of image processing in these systems is to facilitate the extraction of important information from the image contents. Since image processing in evolving imaging technology uses computation- and data-intensive functions, it is therefore performed using a dedicated coprocessor, and the general-purpose processor is freed from this task [1–3]. Today, image processing functions are heavily utilized in portable electronics systems. Such functions therefore demand low-power and area-efficient coprocessors for image processing applications.

The hardware realization of image processing functions has been reported in some existing works [1–3]. In the approaches described [1–3], image processing functions were performed using field-programmable gate arrays (FPGAs). However,

doi:10.1088/978-0-7503-3919-3ch7

these approaches did not design a low-power application-specific processor using the HLS framework. In contrast to the cited works, this chapter presents low-power and area-efficient application-specific reusable IP core designs for image processing functions. The hardware IP cores have been designed as application-specific IP cores using the HLS process to meet the low power requirements of portable systems.

In this paragraph, a detailed description of HLS will be given. HLS, electronic-level synthesis, and algorithmic-level synthesis are significant aspects of the very large-scale integration (VLSI) domain and system-level design. In recent trends, high-level synthesis has attracted a lot of attention in the system-level design and electronics domains and also amongst the vendors catering for computer-aided design (CAD). Numerous vendors of electronic design automation (EDA) working on CAD designs have shifted their attention to electronic-level synthesis [5–7]. High-level synthesis is the transformation of the behavioral description of an algorithm into a format that caters for its register-transfer level (RTL) hardware structure, while satisfying the user-specific design constraints. This is achieved by performing multiparametric optimization at the same time. The vital and noteworthy components of electronic-level synthesis are scheduling, allocation, binding, and design-space exploration. These aspects have attracted a lot of attention from design engineers and researchers in the VLSI and CAD fields, due to its enhanced capability to produce optimized hardware systems designed by HLS. In addition to these aspects, HLS can also be described as the transformation from the description of a behavioral feature to its corresponding hardware description in the form of multiplexers/demultiplexers, storage units, memory elements, and the necessary interconnection elements (called RTL). The significant characteristic here is that the standard and methodical process of high-level synthesis consists of several complex and systematic procedural phases. Significant research and analysis has been conducted and continues to be conducted in the different strata of electronic-level synthesis [8–10].

This paragraph focuses on the elaboration of the several substeps involved for generating an RTL design structure using an HLS framework. As discussed earlier, HLS plays a pivotal role in the design of digital signal processing and image processing hardware cores. Thus, this paragraph provides readers with a deeper understanding of the HLS design methodology used for such multimedia applications. The first step of the HLS framework starts with a library of module information, such as chip area, power dissipation, clock cycles consumed per functional unit, number of functional units available, etc. Furthermore, additional inputs, such as designer-specified constraints, for example, the latency and area/power budget may be provided at this stage. Secondly, the data flow graph (DFG) of the digital signal processing (DSP)/image processing application is also accepted as an input. The DFG is generated either from the transfer function of the application or from its equivalent C code. The next phase is to perform design-space exploration to produce an optimal resource configuration (hardware architecture) based on the library information and orthogonal designer constraints. Design space exploration can be performed using a Nature-inspired algorithm, such as a genetic algorithm, particle swarm optimization, bacterial foraging, etc. [11, 12]. In the next phase, scheduling of the application is performed based on the produced optimal resource configuration (design solution) using one of the scheduling algorithms such as As Soon As Possible (ASAP), force-directed scheduling, etc. Once the scheduling is complete, allocation and binding are performed in an integrated fashion to derive the multiplexer details. Subsequently, the multiplexing scheme is used to find the exact sizes of the multiplexers and demultiplexers involved in the RTL datapath design. Finally, based on the multiplexing scheme, the RTL datapath and controller table is generated for the corresponding DSP/image processing application. The RTL datapath and controller are then implemented in a logic synthesis tool to synthesize and simulate the complete application-specific processor IP design [13–15].

7.2 HLS methodology used to design IP cores for image processing functions

Image processing applications such as blurring, edge detection, magnification, etc. can be performed using a filter kernel to produce a processed image at the output. In order to generate the processed/filtered image, a 2D convolution is performed between the input image and the filter kernel matrix. This section presents a generalization of the convolution function for 3×3 and 5×5 sizes of kernel matrix. In order to design 3×3 and 5×5 filter hardware IP using the HLS process, the convolution (core computation) function is converted into a corresponding DFG. Furthermore, it is fed to the HLS process, which results in the RTL circuit of an image processing filter hardware IP core.

7.2.1 Generalization of convolution for image processing functions

(1) *Generalization of convolution for 3 × 3 filters*

A generic kernel matrix (G) for a 3×3 filter is given by:

$$G = \begin{bmatrix} g_{00} & g_{01} & g_{02} \\ g_{10} & g_{11} & g_{12} \\ g_{20} & g_{21} & g_{22} \end{bmatrix}_{3\times 3} \tag{7.1}$$

where g_{ab} denotes a kernel coefficient. A 2D convolution is computed between the filter kernel and an input image. The approach discussed performs a '*same*' convolution, which results in an output matrix that is the same size as the input matrix. To perform a '*same*' convolution, the size of the input image matrix is first enhanced using a padding rule. This padding rule states that a number 'w' of zero rows and zero columns are added to each side of the original input matrix, where w is equal to the (*size of the kernel matrix* − 1)/2. In the case of a 3×3 convolution filter, the application of this rule results in the addition of one zero row to the top and the bottom and the addition of one zero column to the left and the right of the original input matrix. This results in an increase of two in the input matrix's size for both rows and columns. Thus, the modified input matrix obtained is as follows:

$$I = \begin{bmatrix} 0 & 0 & 0 & \cdots & 0 & 0 \\ 0 & L_{00} & L_{01} & \cdots & L_{0(B-1)} & 0 \\ 0 & L_{10} & L_{11} & \cdots & L_{1(B-1)} & 0 \\ \vdots & \vdots & \vdots & \ddots & \vdots & \vdots \\ 0 & L_{(A-1)0} & L_{(A-1)1} & \cdots & L_{(A-1)(B-1)} & 0 \\ 0 & 0 & 0 & \cdots & 0 & 0 \end{bmatrix}_{N\times M} \tag{7.2}$$

where L_{uv} indicates the pixel values in original input matrix [L] and $A \times B$ is the size of the original input matrix. The values in the modified input matrix are denoted by I_{ij}, and $N \times M$ is the size of the modified input matrix [I] after padding.

Furthermore, a 2D convolution performed between the modified matrix [I] and the kernel [G] results in an output matrix [O] of size $A \times B$ (which is the same as the size of the original input matrix). In general, $A \times B = (N - n + 1) \times (M - m + 1)$, where $n \times m$ is the size of kernel matrix. The generalized image processing function that computes an output pixel O_x is as follows:

$$O_x = \sum_{i,a=min}^{i,a=max} \left(\sum_{j,b=min}^{j,b=max} I_{ij} \times g_{ab} \right) \tag{7.3}$$

where the variable x varies from 0 to $[(N - n + 1) \times (M - m + 1) - 1]$ because the total number of pixels in the output matrix is $(N - n + 1) \times (M - m + 1)$.

In case of 3×3 filters, the variables a and b vary from 0 (the minimum value) to 2 (the maximum value) during the computation of each output value; and the variables i and j vary from 0 (the minimum value) to $N - 1$ (the maximum value for i) and $M - 1$ (the maximum value for j), respectively, throughout the computation of the entire output matrix. Following the computation of each output value, the kernel matrix 'G' is slid over 'I' by one position (it slides one position to the right to compute values in a row and slides one position downward to compute values in a column). Therefore, the minimum and maximum values of j are increased by one after each computation in a row and the maximum value goes up to $M - 1$. Similarly, the minimum and maximum values of i are increased by one after each computation in a column and the maximum value goes up to $N - 1$.

To compute the first pixel value O_0 using a 3×3 filter, the computation function in (7.3) converges as follows:

$$\begin{aligned} O_0 = {} & [(I_{00} \times g_{00}) + (I_{01} \times g_{01}) + (I_{02} \times g_{02})] \\ & + [(I_{10} \times g_{10}) + (I_{11} \times g_{11}) + (I_{12} \times g_{12})] \\ & + [(I_{20} \times g_{20}) + (I_{21} \times g_{21}) + (I_{22} \times g_{22})]. \end{aligned} \tag{7.4}$$

In order to generate a 3×3 filter hardware IP core for image processing functions, the above equation is converted into the corresponding DFG representation and fed to the HLS process. The output of the HLS is the RTL circuit of a 3 × 3 filter hardware IP core.

(2) *Generalization of convolution for 5 × 5 filters*

The generic kernel matrix (G) for a 5 × 5 filter is given by:

$$G = \begin{bmatrix} g_{00} & g_{01} & g_{02} & g_{03} & g_{04} \\ g_{10} & g_{11} & g_{12} & g_{13} & g_{14} \\ g_{20} & g_{21} & g_{22} & g_{23} & g_{24} \\ g_{30} & g_{31} & g_{32} & g_{33} & g_{34} \\ g_{40} & g_{41} & g_{42} & g_{43} & g_{44} \end{bmatrix}_{5\times 5}. \tag{7.5}$$

In the approach discussed, the '*same*' convolution is performed between the 5 × 5 kernel and the input image matrix in the filtering operation. To perform the '*same*' convolution, the size of the input image matrix is first enhanced using the aforementioned padding rule (discussed in section 7.2.1). When the the padding rule is applied to a 5 × 5 convolution filter, two zero rows are added to the top and the bottom, and two zero columns are added to the left and the right of the original input matrix. This results in an increase of four in the size of the input matrix for both rows and columns. The modified input matrix obtained is:

$$I = \begin{bmatrix} 0 & 0 & 0 & 0 & \cdots & 0 & 0 & 0 \\ 0 & 0 & 0 & 0 & \cdots & 0 & 0 & 0 \\ 0 & 0 & L_{00} & L_{01} & \cdots & L_{0(B-1)} & 0 & 0 \\ 0 & 0 & L_{10} & L_{11} & \cdots & L_{1(B-1)} & 0 & 0 \\ \vdots & \vdots & \vdots & \vdots & \ddots & \vdots & \vdots & \vdots \\ 0 & 0 & L_{(A-1)0} & L_{(A-1)1} & \cdots & L_{(A-1)(B-1)} & 0 & 0 \\ 0 & 0 & 0 & 0 & \cdots & 0 & 0 & 0 \\ 0 & 0 & 0 & 0 & \cdots & 0 & 0 & 0 \end{bmatrix}_{N\times M}. \tag{7.6}$$

Furthermore, a 2D convolution performed between the matrix [I] and the kernel [G] of a 5 × 5 filter would result in an output matrix [O]. The generalized image processing function for computing an output pixel O_x is given in (7.3). In the case of a 5 × 5 filter, the variables a and b in (7.3) vary from 0 (the minimum value) to 4 (the maximum value) during the computation of each output value, and the variables i and j in (7.3) vary in the same way, as discussed for 3 × 3 filters (in section 7.2.1).

When we compute the first pixel value O_0 using a 5 × 5 filter, the computation function in (7.3) converges as follows:

$$O_0 = [(I_{00} \times g_{00}) + (I_{01} \times g_{01}) + (I_{02} \times g_{02}) + (I_{03} \times g_{03}) + (I_{04} \times g_{04})]+$$
$$[(I_{10} \times g_{10}) + (I_{11} \times g_{11}) + (I_{12} \times g_{12}) + (I_{13} \times g_{13}) + (I_{14} \times g_{14})]+$$
$$[(I_{20} \times g_{20}) + (I_{21} \times g_{21}) + (I_{22} \times g_{22}) + (I_{23} \times g_{23}) + (I_{24} \times g_{24})]+$$

$$[(I_{30} \times g_{30}) + (I_{31} \times g_{31}) + (I_{32} \times g_{32}) + (I_{33} \times g_{33}) + (I_{34} \times g_{34})] + [(I_{40} \times g_{40}) + (I_{41} \times g_{41}) + (I_{42} \times g_{42}) + (I_{43} \times g_{43}) + (I_{44} \times g_{44})]. \tag{7.7}$$

In order to design a 5 × 5 filter hardware IP for image processing functions, the above equation is converted into the corresponding DFG representation and fed to the HLS process. The output of the HLS is the RTL circuit of a 5 × 5 filter hardware IP core.

7.2.2 Application-specific reusable IP cores for image processing functions

This section presents the design process for application-specific (dedicated) reusable IP cores for the following image processing functions (figure 7.1).

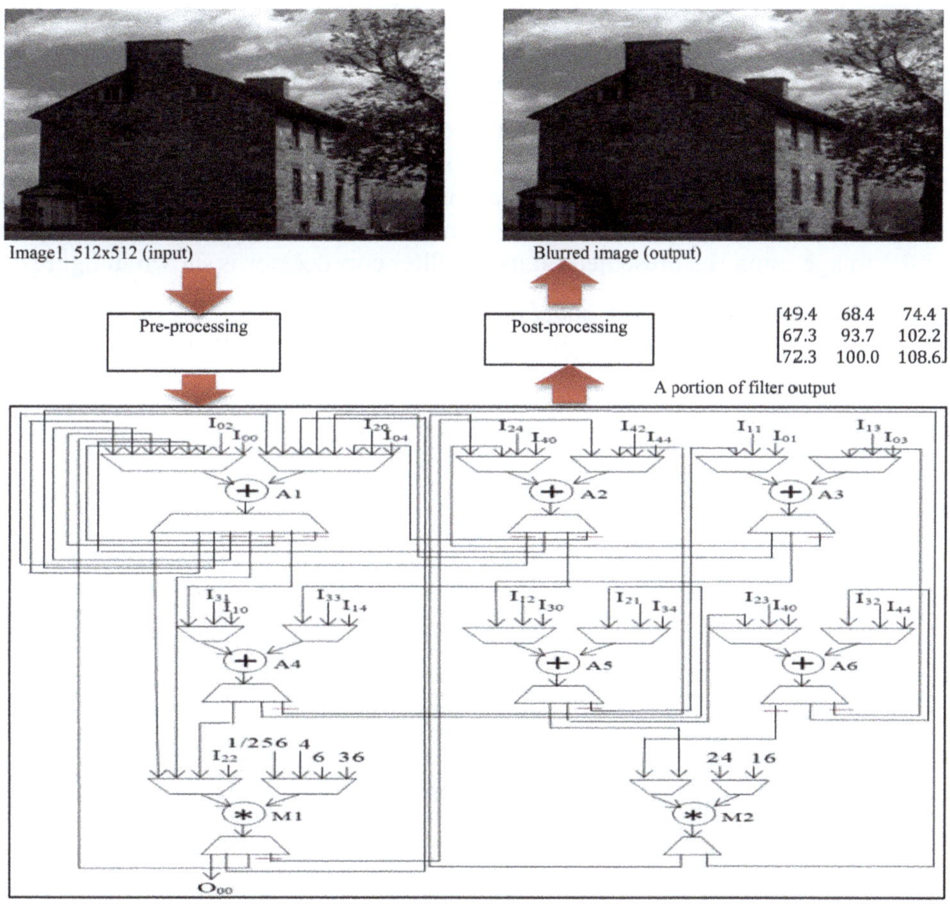

Figure 7.1. End-to-end demonstration of image processing using the presented Gaussian blur filter coprocessor.

(1) *Reusable IP core for a Gaussian blur filter*

The kernel matrix ($[G_{gb}]$) for a Gaussian blur filter with a size of 5×5 is as follows:

$$G_{gb} = \frac{1}{256}\begin{bmatrix} 1 & 4 & 6 & 4 & 1 \\ 4 & 16 & 24 & 16 & 4 \\ 6 & 24 & 36 & 24 & 6 \\ 4 & 16 & 24 & 16 & 4 \\ 1 & 4 & 6 & 4 & 1 \end{bmatrix}_{5\times5}. \tag{7.8}$$

Using this filter kernel, the computation function for a Gaussian blur filter with a size of 5×5 is derived from (7.7), as follows:

$$\begin{aligned} O_0 = \frac{1}{256}[&\{(I_{00} + I_{04} + I_{40} + I_{44}) \times 1\} \\ &+ \{(I_{01} + I_{03} + I_{10} + I_{14} + I_{30} + I_{34} + I_{41} + I_{43}) \times 4\} \\ &+ \{(I_{02} + I_{20} + I_{24} + I_{42}) \times 6\} + \{(I_{11} + I_{13} + I_{31} + I_{33}) \times 16\} \\ &+ \{(I_{12} + I_{21} + I_{23} + I_{32}) \times 24\} + \{(I_{22} \times 36)\}]. \end{aligned} \tag{7.9}$$

The DFG representation of this computation function is fed to the HLS process along with the resource constraints and the module library. The HLS process results in a corresponding RTL circuit for Gaussian filter hardware IP. An end-to-end demonstration of the generation of a blurred image using the presented Gaussian filter coprocessor is shown in figure 7.2.

(2) *Reusable IP core for a Sobel edge detector*

The kernel matrix ($[G_{se}]$) of a Sobel edge detector with a size of 3×3 is as follows:

$$G_{se} = \begin{bmatrix} -1 & 0 & 1 \\ -2 & 0 & 2 \\ -1 & 0 & 1 \end{bmatrix}_{3\times3}. \tag{7.10}$$

Since the kernel matrix is asymmetric, horizontal and vertical flips are applied to the matrix before the convolution process takes place. Hence, the matrix becomes the following:

$$G_{se} = \begin{bmatrix} 1 & 0 & -1 \\ 2 & 0 & -2 \\ 1 & 0 & -1 \end{bmatrix}_{3\times3}. \tag{7.11}$$

Using this filter kernel, the computation function of a Sobel edge detector is derived from (7.4), as follows:

$$O_0 = [(I_{00} - I_{02})] + 2_*[I_{10} - I_{12}] + [(I_{20} - I_{22})]. \tag{7.12}$$

Using the HLS process, an RTL circuit for the presented Sobel edge detector hardware IP core is designed using the computation function. An end-to-end

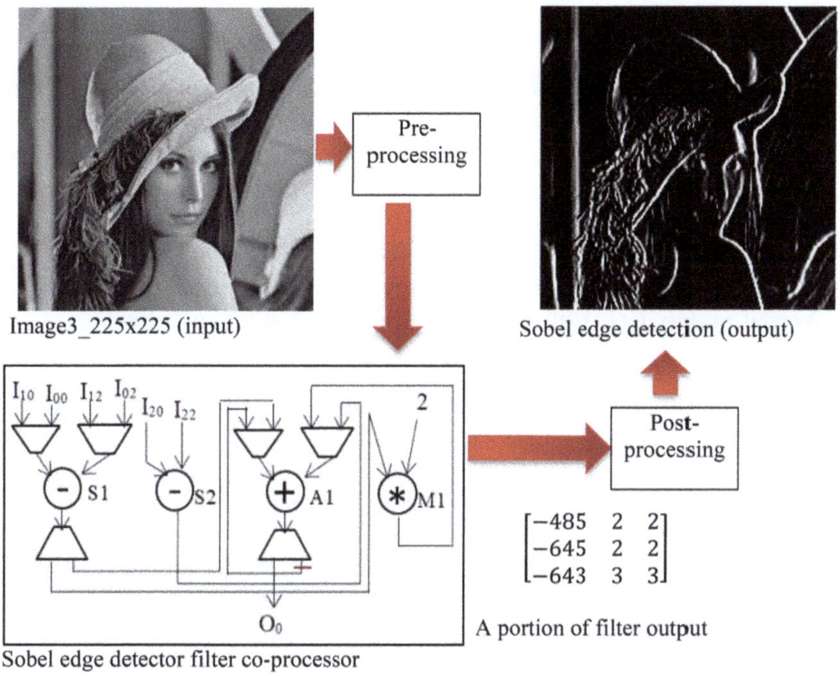

Figure 7.2. End-to-end demonstration of image processing using the presented Sobel edge detector filter coprocessor.

demonstration of the generation of an output image using the presented Sobel edge detector hardware IP core is shown in figure 7.2.

(3) *Reusable IP core for a Prewitt gradient edge detector*

The kernel matrix ($[G_{pe}]$) of a Prewitt gradient edge detector with a size of 3 × 3 is as follows:

$$G_{pe} = \begin{bmatrix} -1 & 0 & 1 \\ -1 & 0 & 1 \\ -1 & 0 & 1 \end{bmatrix}_{3\times 3}. \tag{7.13}$$

Since the kernel matrix is asymmetric, horizontal and vertical flips are applied to the matrix before the convolution process takes place. Hence, the matrix becomes the following:

$$G_{pe} = \begin{bmatrix} 1 & 0 & -1 \\ 1 & 0 & -1 \\ 1 & 0 & -1 \end{bmatrix}_{3\times 3}. \tag{7.14}$$

Using this filter kernel, the computation function of a Prewitt edge detector is derived from (7.4) and given as follows:

$$O_0 = [(I_{00} + I_{10} + I_{20})] - \left[(I_{02} + I_{12} + I_{22})\right]. \tag{7.15}$$

Using the HLS process, an RTL circuit for a Prewitt edge detector hardware IP core is designed using the computation function. An end-to-end demonstration of the generation of an output image using the presented Prewitt edge detector hardware IP core is shown in figure 7.3.

(4) *Reusable IP core for image magnification*

To achieve image magnification using the convolution approach, the kernel matrix ($[G_z]$) of a magnification filter is given below:

$$G_z = \begin{bmatrix} 0.25 & 0.5 & 0.25 \\ 0.5 & 1 & 0.5 \\ 0.25 & 0.5 & 0.25 \end{bmatrix}_{3\times3}. \tag{7.16}$$

Before applying the filter kernel to the input image for convolution, zero rows and zero columns are added in between the rows and columns and at the border of the input matrix. Thus, the modified input matrix used for image magnification is shown below:

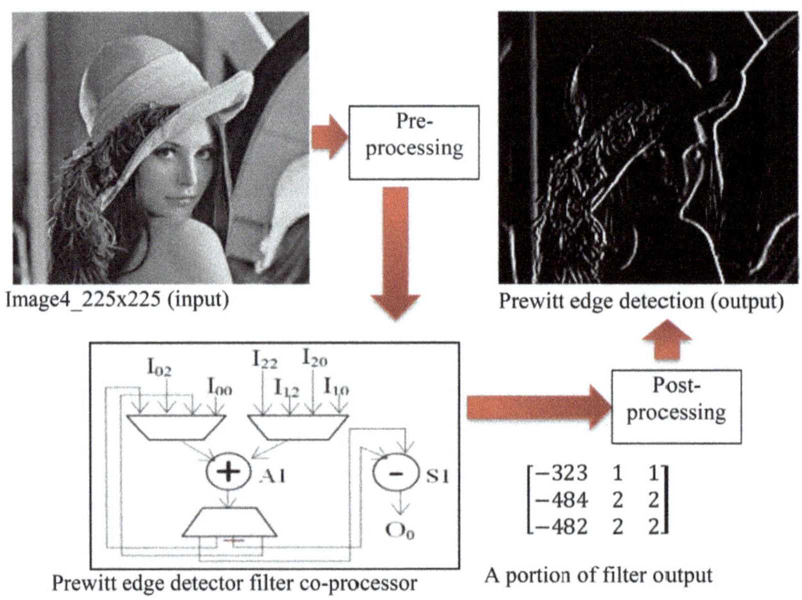

Figure 7.3. End-to-end demonstration of image processing using the presented Prewitt edge detector filter coprocessor.

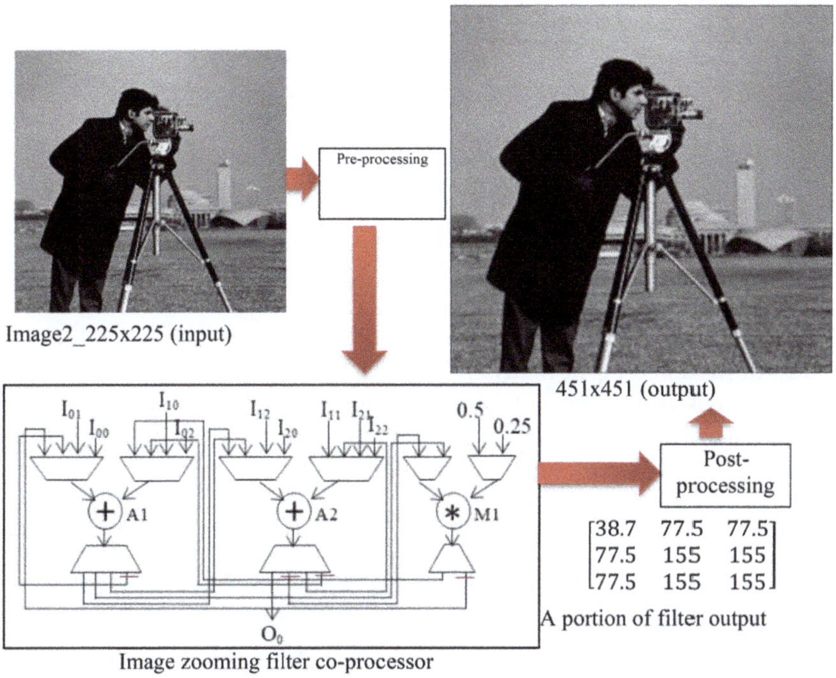

Figure 7.4. End-to-end demonstration of image processing using the presented magnification filter coprocessor.

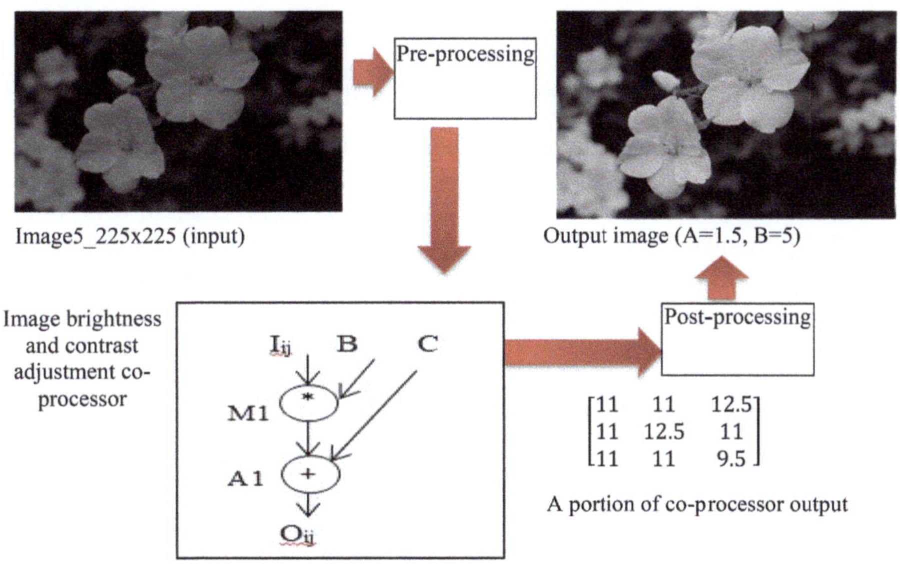

Figure 7.5. End-to-end demonstration of image processing using presented brightness and contrast adjustment coprocessor.

$$Z = \begin{bmatrix} 0 & 0 & 0 & 0 & 0 & \cdots & 0 & 0 \\ 0 & L_{00} & 0 & L_{01} & 0 & \cdots & L_{0(B-1)} & 0 \\ 0 & 0 & 0 & 0 & 0 & \cdots & 0 & 0 \\ 0 & L_{10} & 0 & L_{11} & 0 & \cdots & L_{1(B-1)} & 0 \\ 0 & 0 & 0 & 0 & 0 & \cdots & 0 & 0 \\ \vdots & \vdots & \vdots & \vdots & \vdots & \ddots & \vdots & \vdots \\ 0 & L_{(A-1)0} & 0 & L_{(A-1)1} & 0 & \cdots & L_{(A-1)(B-1)} & 0 \\ 0 & 0 & 0 & 0 & 0 & \cdots & 0 & 0 \end{bmatrix}_{(2A+1)\times(2B+1)}$$

where $A \times B$ is the size of original input matrix and $(2A + 1) \times (2B + 1)$ is the size of modified input matrix [Z] after it is filled with zero rows and columns. The matrix [I] (shown in (7.2)) is now derived from matrix [Z] by applying a padding rule to matrix [Z] for 3×3 filters. Using the magnification filter kernel shown in (7.16), the computation function is derived (from (7.4)), as follows:

$$O_0 = [(I_{00} + I_{02} + I_{20} + I_{22}) \times 0.25] + \left[(I_{01} + I_{10} + I_{12} + I_{21}) \times 0.5\right] + I_{11} \quad (7.17)$$

where I_{ij} indicates the values in the 'zero-padded, alternately zero filled' matrix [I]. Using the HLS process, an RTL circuit for the presented image magnification filter hardware IP core is designed using the computation function shown in (7.17). An end-to-end demonstration of the generation of an output magnified image using the presented image magnification filter hardware IP core is shown in figure 7.4.

(5) *Reusable IP core for image contrast and brightness adjustment*

The computation function for image contrast and brightness adjustment is as follows:

$$O_{uv} = [((L_{uv*}B) + C)] \quad (7.18)$$

where L_{uv} and O_{uv} indicate the pixel values in the original input matrix and the output matrix, respectively. 'B' and 'C' are the brightness and contrast adjustment factors, respectively. Using the HLS process, an RTL circuit of the presented image brightness and contrast adjustment hardware IP is designed using the computation function. An end-to-end demonstration of the generation of an output image using the presented brightness and contrast adjustment hardware IP is shown in figure 7.5.

7.3 Case studies and analysis

This section presents case studies and an analysis of various image processing applications, viz. the Gaussian blur filter, the Sobel edge detector, the Prewitt gradient edge detector, image magnification, and image contrast/brightness adjustment. Applications such as blurring, edge detection, and image magnification are executed using dedicated convolution filter hardware, while image contrast and brightness adjustment are executed through non-convolution hardware.

The chapter has presented several hardware designs (IP cores) for the above-mentioned applications. The presented application-specific reusable IP cores for image processing applications have been analyzed in terms of gate count and power. The gate count and power have been computed using a 15 nm-technology-based open-source library [4].

7.3.1 Gate-count analysis

The gate counts of the presented hardware IP cores for image processing functions have been analyzed for two versions, viz. the baseline (non-optimized) and the optimized version. Here, 'baseline' denotes the IP cores designed using chosen resource constraints. In order to achieve a lower gate count, optimized versions of the presented IP cores have been designed using optimal resource constraints (generated by the design-space exploration process). Table 7.1 presents a comparison of the gate counts of the baseline and optimized versions of the hardware IP cores. As shown in the table, a significant reduction in gate count is achieved.

7.3.2 Power analysis

The power consumption of the presented hardware IP cores is shown in table 7.2. As shown in the table, the presented IP cores consume small amounts of power. This is because the presented IP cores have been designed as application-specific processors using the HLS framework, which offers the flexibility to optimize the area and the power consumption. The percentage reduction in the power of the presented

Table 7.1. Gate-count (GC) analysis of the presented image processing hardware IP cores.

Benchmark	Baseline GC	Optimized GC
Gaussian blur	11 360	10 752
Magnification	4016	3456
Sobel edge detector	2592	2464
Prewitt edge detector	1872	1760
Brightness and contrast	1120	1120

Table 7.2. Power analysis of image processing hardware IP cores.

Benchmark	Baseline (μW)	Optimized (μW)
Gaussian blur	90.91	86.05
Magnification	32.14	27.66
Sobel edge detector	20.74	19.72
Prewitt edge detector	15.00	14.08
Brightness and contrast	8.96	8.96

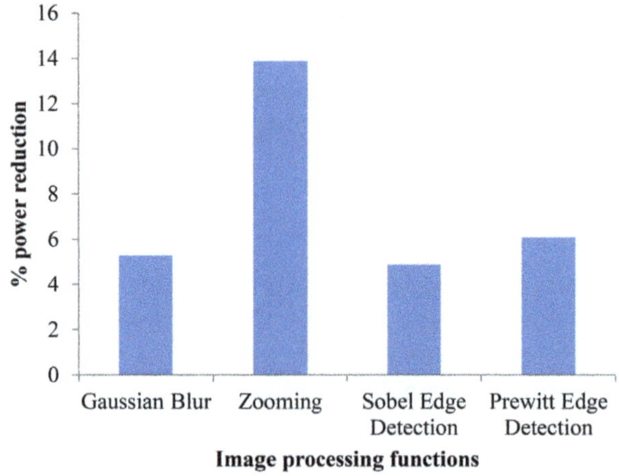

Figure 7.6. Percentage reduction in power due to optimal resource constraints.

optimized IP core versions compared to the IP core baseline versions is shown in figure 7.6. The above analysis highlights the low power requirements of the presented hardware IP cores for image processing functions.

7.4 Conclusions

This chapter presented a discussion of a low-power design of dedicated hardware IP cores for image processing functions, such as Gaussian blur, magnification, Sobel/ Prewitt gradient edge detection, and brightness and contrast adjustment. Some of the salient features of this chapter are the following:

- A discussion of the background of HLS for VLSI design.
- A discussion of the reason that HLS is significant for DSP and image processing applications.
- The subphases involved in HLS design methodology for RTL processor generation.
- The methodology used to perform hardware-based convolution for image processing functions, such as blurring, edge detection, and magnification.
- A presentation of the generalization functions for a 3×3 filter kernel.
- A presentation of the pixel computation functions for 3×3 convolution filter kernel.
- A presentation of the generalization functions for a 5×5 filter kernel.
- A presentation of the pixel computation functions for a 5×5 convolution filter kernel.
- Design of application-specific reusable IP cores for image processing functions, such as a Gaussian blur filter, a Sobel edge detector, a Prewitt gradient edge detector, image magnification, and image contrast/brightness adjustment.
- A presentation of case studies, including power and gate count analyses for various image processing functions.

Acknowledgements

The authors wish to acknowledge the assistance of Mahendra Rathor, PhD candidate at Indian Institute of Technology, Indore.

Bibliography

[1] Ortega-Cisneros S *et al* 2014 Real time hardware accelerator for image filtering *Progress in Pattern Recognition, Image Analysis, Computer Vision, and Applications CIARP 2014: Progress in Pattern Recognition, Image Analysis, Computer Vision, and Applications* vol 8827 pp 80–7

[2] Vourvoulakis J V, Lygouras J and Kalomiros J A 2012 Acceleration of image processing algorithms using minimal resources of custom reconfigurable hardware *16th Panhellenic Conf. on Informatics* pp 68–73

[3] Tsiktsiris D, Ziouzios D and Dasygenis M 2018 A portable image processing accelerator using FPGA *7th Int. Conf. on Modern Circuits and Systems Technologies (MOCAST)* pp 1–4

[4] *NanGate 15 nm Open Cell Library* http://nangate.com/?pageid=2328

[5] McFarland M C, Parker A C and Camposano R 1990 The high-level synthesis of digital systems *Proc. IEEE* **78** 301–18

[6] McFarland M C, Parker A C and Camposano R 1988 Tutorial on high level synthesis *Proc. of the 25th ACM/IEEE Design Automation Conf.* (Washington, DC: IEEE Computer Society Press) pp 330–6

[7] Gallagher J C, Vigraham S and Kramer G 2004 A family of compact genetic algorithms for intrinsic evolvable hardware *IEEE Trans. Evolution. Comput.* **8** 111–26

[8] Bailey B, Martin G and Piziali A 2007 *ESL Design and Verification: A Prescription for Electronic System Level Methodology* (San Francisco, CA: Morgan Kaufmann)

[9] Williams A C, Brown A D and Zwolinski M 2000 Simultaneous optimisation of dynamic power, area and delay in behavioural synthesis *IEE Proc. Comput. Digit. Tech.* **147** 383–90

[10] Zhang X and Ng K W 2000 A review of high-level synthesis for dynamically reconfigurable FPGAs *Microprocess. Microsyst.* **24** 199–211

[11] Ascia G, Catania V, Di Nuovo A G, Palesi M and Patti D 2007 Efficient design space exploration for application specific systems-on-a-chip *J. Syst. Archit.* **53** 733–50

[12] Haubelt C and Teich J 2003 Accelerating design space exploration using pareto-front arithmetics *Proc. of Asia and South Pacific Design Automation Conf. (ASP-DAC'03)* pp 525–31

[13] Coussy P, Gajski D D, Meredith M and Takach A 2009 An introduction to high-level synthesis *IEEE Des. Test Comput.* **26** 8–17

[14] Ahmad I, Dhodhi M K and Chen C Y R January 1995 Integrated scheduling, allocation and module selection for design-space exploration in high-level synthesis *IEE Proc. Comput. Digit. Tech* **142** 65–71

[15] Elliot J P 1999 *Understanding Behavioral Synthesis: A Practical Guide to High-level Design* (Boston, MA: Springer)

IOP Publishing

Advances in Image and Data Processing using VLSI Design, Volume 1

Smart vision systems

Sandeep Saini, Kusum Lata, Abhishek Sharma and G R Sinha

Chapter 8

An efficient underwater image cryptosystem that uses a novel hybrid algorithm

Alpana Sharma, Venkatadri Marriboyina and Kamlesh Gupta

In this chapter, we propose a hybrid underwater image cryptosystem that combines an image enhancement algorithm with the secure force algorithm (which uses 96-bit symmetric key generation). The system is specially developed for distorted underwater images to enhance visual quality, dynamically remove interference, and furthermore, to protect the encrypted image from third-party attacks and promote secure data sharing in underwater communications. The proposed cryptosystem is implemented through a hybrid approach by combining the discrete cosine transform (DCT) and a tuned tri-threshold fuzzy intensification operator (TTFIO). The background of the underwater image is detected by DCT, and the visual quality is improved using TTFIO and Weber's law. This approach enhances image contrast and dynamically removes interference. Finally, the proposed algorithm is tested on various underwater images collected from the internet and compared with the original block-based scheme. Simulation experiments show that the proposed algorithm provides better contrast enhancement and reconstruction of the underwater images.

8.1 Introduction

In the digital world, the available modes of communication have grown exponentially beyond anything that could have been imagined. Advances in communication technologies, the invention of sensors in the Internet of Things (IoT), and the technological improvements in computing devices have led to increasing digital data transmission. Digital information not only comprises text data but also a large volume of multimedia data (video, audio, and images), which are comparatively more sizable than textual information alone [1–3]. Certain sectors require secure communications, especially for underwater object identification in military

doi:10.1088/978-0-7503-3919-3ch8

communications, image transfers in healthcare data communications, the enhancement of security in data transfers in the machine-to-machine communication of images, the Internet of Things (IoT), and digital image communications in smart infrastructure development. Security is essential for secure communications between digital devices on a public network [4–7]. All data-sensitive applications require fast and secure digital data communications with better image quality for object identification and information transmission [8, 9]. Image enhancement is the process of detecting background information and increasing image visibility [10]. It is hard to discern objects clearly underwater, due to the presence of floating particles in the physical medium [11]. Image enhancement plays a vital role in military communications [12], the positioning of shipping [13], the positioning of underwater mining, and in underwater surveillance systems [14]. Oceanic exploration is another area of interest, particularly in searches for ancient shipwrecks [15]. Research has identified issues with the sonar images used to locate shallow-water fish and wrecks: they suffer from noise and lack image clarity [16, 17]. Sonar image systems suffer from poor resolution when used for both long- and short-distance target detection [18]. Most conventional techniques are unable to provide better image enhancement; therefore, there is a requirement to design and implement advanced underwater image enhancement techniques [19].

In the digital world, most applications use digital transmission to provide quality services in the military, medical, and surveillance fields. Digital transmission supports reliable communication services, such as video conferencing and bar code scanner and cradle transmission. In these sectors, images play a vital role to in the detection and exchange of information across the globe. However, the transmission of digital images through public networks causes new problems with security and privacy. Therefore, this has resulted in the development of a new platform called underwater image transmission. This provides adequate transmission of data over long distances, but its related security issues must be addressed carefully. Thus, cryptography techniques are highly recommended to promote secure transmission, but these need an efficient encryption and decryption strategy. In underwater communications, images are transmitted across public networks to other devices, but unauthorized users may attempt to access the information. So, security layers should be superimposed on the transport protocol, providing privacy protection, information integrity, and also protection from unauthorized access. The underlying concept is to change the content's texture or plaintext into a random-looking texture, also known as the ciphertext of an original image. This is referred to as a discrete-valued cryptosystem [20]. Hence, an unauthorized receiver will not understand the content of an image and it will be recognizable only to an authorized recipient. The content of an original image is then reconstructed with help of security cipher key generation using the discrete-valued cryptosystem [21]. The digital transfer of multimedia data faces security issues over underwater communications, which can be addressed by an efficient hybrid secure image enhancement algorithm.

In this paper, we propose a hybrid underwater image cryptosystem which combines the proposed image enhancement (dynamic histogram enhancement technique (DHET)) [22] and the secure force algorithm (96-bit symmetric key generation) [23]. This cryptosystem is specially developed for distorted underwater images; it performs visual quality restoration, dynamic removal of interference, and proper reconstruction of the enhanced image. It also protects the encrypted image from third-party attacks and promotes secure data sharing between vehicle nodes. Figure 8.1 shows the framework of the proposed hybrid underwater image cryptosystem. It combines DCT and a tuned tri-threshold fuzzy intensification operator, and provides better contrast enhancement of the underwater images. First, the background of the underwater image is detected by DCT and the visual quality is improved using the tuned tri-threshold fuzzy intensification operator and Weber's law, thus enhancing image contrast and dynamically removing interference. Finally, the proposed algorithm is tested on various underwater images collected from internet and compared with the original block-based scheme. The results of simulations and experiments show that the proposed algorithm provides better contrast enhancement and reconstruction of the underwater images. This chapter is organized as follows: first, the dynamic histogram enhancement technique is discussed in section 8.2. The secure force algorithm based on a 96-bit symmetric key for underwater image security is described in section 8.3. The evaluation parameters for the hybrid underwater image cryptosystem are discussed in section 8.4. The experimental evaluation is carried out using underwater seabed images collected from internet and compared with the original 32-bit and 64-bit schemes in section 8.5, and finally, the conclusions and the scope of future research are addressed in section 8.6.

8.2 Dynamic histogram enhancement technique

In this section, a DHET is discussed in the context of distorted underwater images, for which it performs visual quality restoration, dynamic removal of interference, and proper reconstruction of enhanced images. It uses a combination of a DCT and a TTFIO to perform better contrast enhancement of underwater images [22]. First, the background of the underwater image is detected by DCT, and the visual quality is improved using a TTFIO and Weber's law, thus enhancing the image contrast and dynamically removing interference. The DHET consists of three different phases,

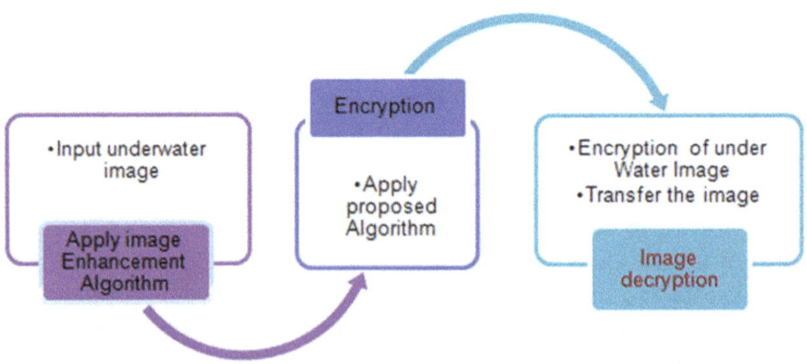

Figure 8.1. Proposed framework for the hybrid underwater image cryptosystem.

known as the block phase, the DCT phase, and the tuned tri-threshold intensification operator phase. In the first phase, an input image (RGB) is converted into a YCbCr image and then divided into number of blocks. This gives more attention to each block, improving the image acquisition system. In the second phase, the background features of each block are extracted, and DCT builds the significant object details by combining all the blocks into a single image. In the third phase, a tuned tri-threshold intensification operator is used to enhance the intensity of the color image. This is made possible by the use of a primary fine-tuning parameter (zeta) which minimizes the color contrast problem and manages the image color accuracy. The processed image is then split into R, G, and B channels. The TTFIO technique is used to improve the intensity of the color image. This technique starts with the first tuning parameter, i.e. zeta. The processed image is then decomposed into R, G, and B channels. The membership function's output is analyzed for the red, green, and blue channels. The pixel intensity is uniformly distributed in each channel and then combined into a single image, using which, an evaluation of the peak signal-to-noise ratio (PSNR), the mean squared error (MSE), horizontal and vertical correlations, and entropy is performed. Figure 8.2 shows a flowchart of the proposed image enhancement algorithm.

The pseudocode of the proposed algorithm is as follows:

Parameter Initialization

#1 Get the input color image (preferably in Joint Photographic Experts Group (JPEG) format)

#2 Convert the color image into a grey scale image (i.e. convert RGB into YCbCr)

#3 Set the block size (the preferred block size is 8 × 8)

#4 Convert the image into a number of blocks for interference removal

#5 for iteration starts<=last block

#6 Adjust the background illustration DCT

#7 end for

#8 Merge all the blocks of the image

#9 Convert the grey scale image into a color image (i.e. convert YCbCr into RGB)

#10 Set zeta = 0.5

#11 Split the image into R, G, and B components

#12 Set threshold values

#13 Compute membership functions for every component intensification operator

#14 if, True:, False:

#15 Perform histogram equalization of the R component

#16 if, True:, False:

#17 Perform histogram equalization of the G component

#18 if, True:, False:

#19 Perform histogram equalization of the B component

#20 Concatenate the R, G, and B components to make a single processed RGB image

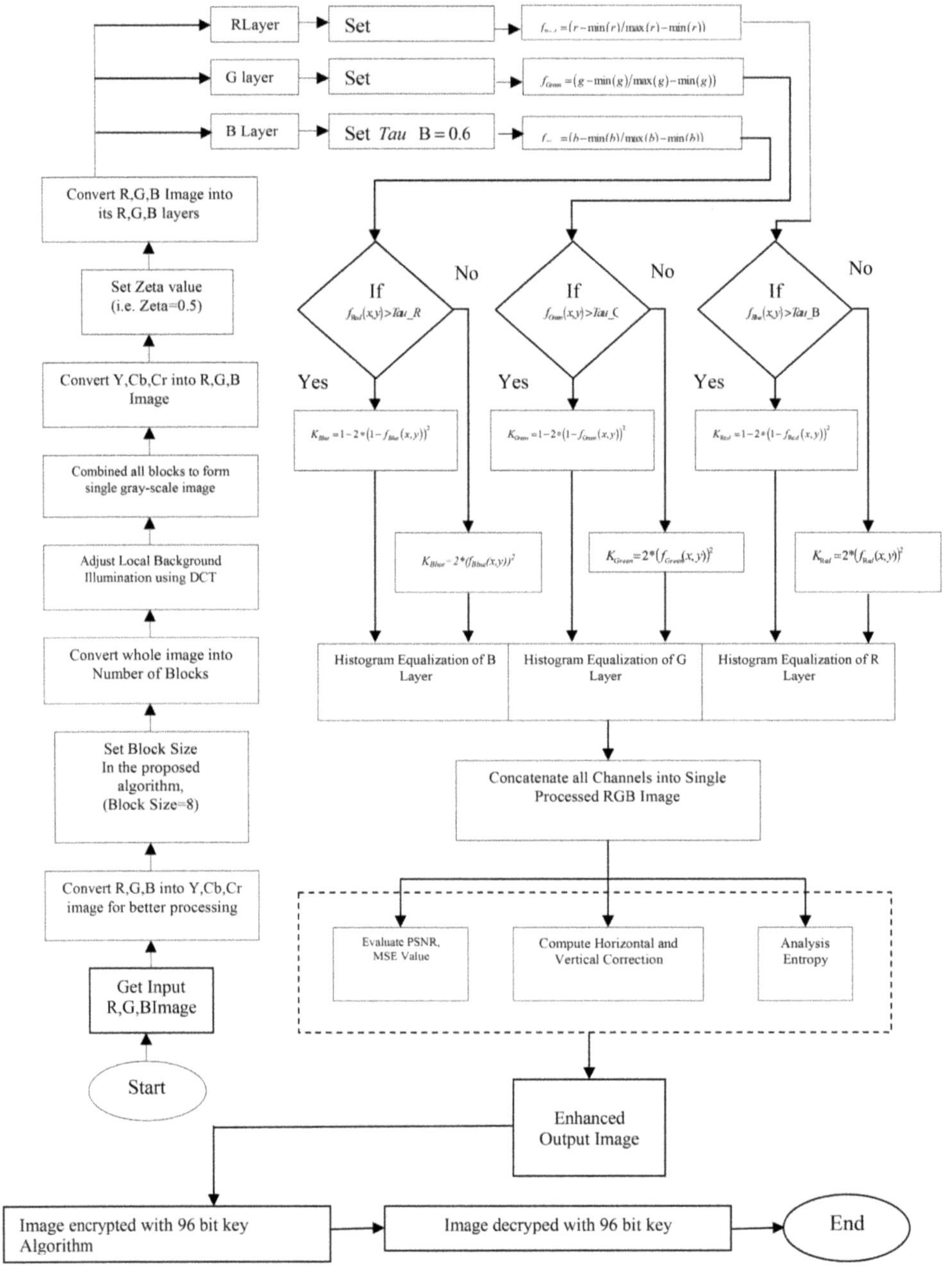

Figure 8.2. Flowchart of the dynamic histogram enhancement algorithm.

#21 Evaluate PSNR and MSE
#22 Compute the horizontal and vertical correlations
#23 Analyze the entropy
#24 Enhance the output image

#25 end if
#26 end if
#27 end if

8.3 Improved secure force algorithm for underwater image transmission

This section describes the framework of a new secure force algorithm based on a 96-bit symmetric key for underwater image security. This is a kind of a symmetric cryptographic algorithm, which uses basic arithmetic operations to perform encryption and decryption. Hence, it uses a Feistel structure to analyse the security inefficiencies associated with vehicle image capture systems. It mainly involves the logical operations (AND, OR, XOR, and XNOR), a swapping process, and the left and right shift functions. Thus, it causes minimum interoperability issues with the encoder and brings optimized image encryption and the benefits of secure transmission to operations deep underwater. In addition to this, various permutation and substitution techniques are used to improve the security factors of the underwater transmission system. This means that it protects the encrypted image from third-party attacks and promotes secure data sharing between vehicle nodes. Several parameters, such as NPCR, UACI, MAE, and correlation coefficients are used to evaluate the degree of security enhancement provided by the underwater transmission system. The simulation results confirm that the accuracy and efficiency of the proposed secured force algorithm are improved, compared with the standard existing security algorithm (with 32-bit and 64-bit keys). A diagonal correlation and histogram analysis is included in the evaluation process to monitor the accuracy and efficiency of the proposed secure force algorithm. This algorithm consists of four main phases, known as the key-generation block, the key management protocol, the encryption block and the decryption block. Hence, it is fully influenced by Feistel structure and uses basic bit-changing functions (shifting and swapping) and logical functions (AND, OR, XOR, and XNOR). Thus, it improves the encryption portion of the secured force algorithm by providing an optimum solution for security threats and generating cipertext through optimized data sets.

8.3.1 Key-generation block

The first phase of the secure force algorithm is the key-generation block [23]. This block generates different sizes of encryption and decryption key using key-expansion facilities based on the input cipher length. Thus, it reduces weak key generation and increases key strength. It contains two main mechanisms: (i) key expansion and (ii) round key selection. Key expansion is performed using a fixed matrix (FM) which executes the logical operations (XOR and XNOR), the circular shift (CS) operation, and matrix multiplication. The security factors are improved through permutation and substitution techniques and separate tables (P- and T-tables) are maintained to update the statuses of the correlation coefficients.

The key bits are left or right shifted to generate the cipher key in order to maintain a continuous generation of keys for the incoming input bits of the image. Thus, round keys (Kr) are the result of starting the input cipher key by using a key plan. The key-generation phase of the secure force algorithm is described in step-by-step detail.

Step1: Generate a 96-bit cipher key (K) with six separate blocks consisting of 16 bits.

Step2: Each block of 16 bits is rearranged into matrix row-wise by repeated left-shift (LS) operations.

Step3: Similarly, we generate a matrix column-wise by repeated logic operations (XOR & XNOR).

Step4: The matrices are rearranged back into the form of a 96-bit array structure.

Step5: A stream of 96 bits is sent to the P-table, where a matrix is generated row-wise by repeating LS operations.

Step6: Then, the FM is multiplied by the matrix generated in step 5, and the resulting 16-bit matrix is transformed into a 96-bit stream.

Step7: We again apply LS to the resulting 96-bit stream and convert it into a 16-bit matrix block column-wise by repeated logic operations (XOR and XNOR).

Step8: Furthermore, the 16-bit block of the matrix column is divided into a 4-bit stream and passed through an XOR operation to obtain a 4-bit key.

Step9: The 4-bit key is used in the permutation and substitution method to generate a 16-bit block just before making six subkeys (K1, K2, K3, K4, K5, K6) of 16 bits each.

8.3.2 Key management block

Secure key sharing is established for the encoder using the localized encryption and authentication protocol (LEAP) [20].

8.3.3 Encryption block

In the third phase of the secure force algorithm, the encryption block is initiated once the key generation has started, which influences the secure receipt of the key-expansion block through the encoder from the LEAP protocol. Here, the encryption procedure is an easy operation, being the sum of AND, OR, XOR, XNOR, left shift (LS), substitution (S boxes), and swapping operations, which are performed in a direction that generates obfuscation and diffusion. The following steps show the encryption process.

Step1: The 64 bits of input plaintext (X) are divided into two halves of 32 bits each.

Step2: Again, each 32-bit half is subdivided into two further halves of 16 bits.

Step3: During each round, a 16-bit block is exchanged. The reason behind for this is to alter the original location of the information to make a more difficult ciphertext. The corresponding subkeys (K1, K2, K3, K4, K5, K6) are XNORed by the left and right halves of each round.

Step4: The outcome of each round is fed as an input to the subsequent round and mapped using the F function (given by the equation below); the F function uses substitution (S boxes), AND, OR, and the left-shift (LS) operation.

$$F = \mathrm{OR}(\mathrm{S} - \mathrm{bo}x\mathrm{es}(\mathrm{AND}(\mathrm{LS}(16\ \mathrm{bits}/4)))) \tag{8.1}$$

Step5: The output of the F function is then XORed by swapping it with 16 bits of a similar round; the result of this is the obfuscation of the data. This completes the encryption process. The decryption procedure just consists of the process described at the top of figure 8.2.

8.4 Evaluation parameters used for the hybrid underwater image cryptosystem

This section explores the parameters involved in the analysis of the processing efficiency of the proposed algorithm.

8.4.1 Peak signal-to-noise ratio and mean squared error

The image enhancement is intended to decrease the noise level in the digital image. The peak signal-to-noise ratio and mean squared error (MSE) are used to evaluate the effectiveness of the processing. Image enhancement methods must not increase the noise level as a result of the enhancement process [15]. The features of the enhanced image are evaluated subjectively as well as objectively, in terms of their visual demand and the incidence of unnecessary color artifacts. The metrics used are the PSNR) and the (MSE) [24].

Mathematically, the individual quantitative measurements can be given as follows [15]:

$$\text{PSNR} = 10\log_{10}\frac{255^2}{\text{MSE}} \tag{8.2}$$

where

$$\text{MSE} = \frac{\sum_{i-1}^{w}\sum_{j-i}^{H}(x_{i,j} - x\bar{}_{i,j})}{W \times H} \tag{8.3}$$

and $x_{i,j}$ represents the original pixel, $x\bar{}_{i,j}$ is used for the reconstructed pixel, and the size of the entire image is $W \times H$.

If the PSNR value is higher, then the visual quality and construction of objects are improved in the enhanced image.

8.4.2 Mean absolute error

The mean absolute error (MAE) measures the degree of modification between the encrypted image E and the source image P. Let W and H be the width and height of the source image, respectively; then,

$$\text{MAE} = \frac{1}{W \times H}\sum_{i=1}^{H}\sum_{j=1}^{W}p(i, j) - E(i, j). \tag{8.4}$$

Here, (i, j) is the original pixel assessment at the position (i, j), and (i, j) is also the encrypted pixel assessment at the same place [9]. The mean absolute error is given by the MAE. Larger values of MAE are caused by improved image security [7].

8.4.3 Horizontal and vertical correlations

A correlation coefficient [25, 26] determines the dissimilarity between two variables. It predicts the change within a single variable with reference to the other variable. Usually, correlation coefficient values take values up to +1; +1 denotes a positive correlation, while −1 refers to a negative correlation. The horizontal and vertical correlations show how powerfully pairs of variables are related. To calculate the correlation between two different horizontally and vertically neighbouring pixels in the enhanced image and the original image, respectively, the following two formulas are used:

$$\mathrm{cov}(x, y) = E(x - E(x))(y - E(y)) \tag{8.5}$$

$$r_{x,y} = \frac{\mathrm{cov}(x, y)}{\sqrt{D(x)}\sqrt{D(y)}} \tag{8.6}$$

where x and y are the values of two neighbouring pixels inside the image. During the mathematical computation, discrete formulas are used, as given below:

$$E(x) = \frac{1}{N}\sum_{i=1}^{N} x_i \tag{8.7}$$

$$D(x) = \frac{1}{N}\sum_{i=1}^{N}(x_i - E(x))(y - E(y)) \tag{8.8}$$

$$\mathrm{cov}(x, y) = \frac{1}{N}\sum_{i=1}^{N}(x_i - E(x))(y_i - E(y)). \tag{8.9}$$

8.4.4 Number of pixels change rate (NPCR) and unified average change intensity (UACI)

An attractive characteristic of an encrypted image is that it is responsive to small changes of its corresponding plain image (e.g. a change of just a single pixel). A challenge would be able to create a small change inside the input image with the intention of viewing the change in the output. Through this technique, it might be possible to establish a significant connection between the original image and the cipher image. However, one little change in the plain image might cause an important modification of the cipher image because the values are subjected to diffusion and confusion, in which case, the differential attack would not really be able to achieve effectiveness and would become relatively ineffective. The NPCR calculates the percentage representing the ratio of the number of different pixels to the total number of pixels within the two images. The UACI calculates the average intensity of the difference between the two images [27]. High values of the NPCR and the UACI denote improved encryption. Consider two cipher images, C1 and C2, for which the equivalent plain image now has one pixel altered. The NPCR of those two images is described by:

$$\mathrm{NPCR} = \frac{\sum_{i,j} D(i, j)}{W \times H} \times 100\%. \tag{8.10}$$

Here, W and H are the width and the height of the image, and $D(i, j)$ can be defined as:

$$D(i, j) = \begin{cases} 0, & \text{if } \mathrm{C1}(i, j) = \mathrm{C2}(i, j) \\ 1, & \text{if } \mathrm{C1}(i, j) \neq \mathrm{C2}(i, j) \end{cases}. \tag{8.11}$$

An additional metric, the UACI, is defined by the following formula:

$$\mathrm{UACI} = \frac{1}{W \times H} \times \sum_{i,j} \left[\frac{\mathrm{C1}(i, j) - \mathrm{C2}(i, j)}{255} \right] \times 100\%. \tag{8.12}$$

A strong image cryptosystem must possess high sensitivity to minor changes in the pending plain image, in order to qualify as offering high security. The NPCR and UACI are two basic metrics used to measure the encrypted image's resistance to differential attacks.

8.4.5 Encryption and decryption times

These are parameters used to calculate the computational overhead of the cryptography algorithms. The effectiveness of the algorithm is calculated using the quantity of time required to encrypt an image. During such an analysis, the real time spent on CPU cycles is used when calculating the execution time. This is significant, as the aim of the encryption and decryption algorithms is to be sufficiently quick to meet the requirements [28]. The designer must attempt to optimize a cryptosystem to make the execution time as small as possible. The encryption time is the time required to change plaintext to ciphertext. The encryption time depends on the key size, the plaintext block size, and the mode. The encryption time is calculated in seconds. The encryption time impacts the system performance. Less time spent here makes the system quick as well as usable. The decryption time is the time taken to obtain plaintext from the ciphertext. The decryption time is similarly chosen to be lower, similarly to encryption time, to make a system that is usable as well as quick. The decryption time also impacts the system performance. The decryption time is calculated in seconds [26].

8.5 Results and discussion

This section explores the qualitative and quantitative evaluation of the proposed hybrid underwater image cryptosystem which combines image enhancement (DHET) and the 96-bit symmetric key secured force algorithm, which is comprehensively compared with the standard 32-bit and 64-bit methods. Figure 8.3(a) shows the test image dataset for the hybrid underwater image cryptosystem. The experimental results include two major parts. Part A discusses the qualitative evaluation of the encryption process for different sizes (256 × 256) and (512 × 512) of test image.

Figure 8.3(a). Test image dataset used in the experiments.

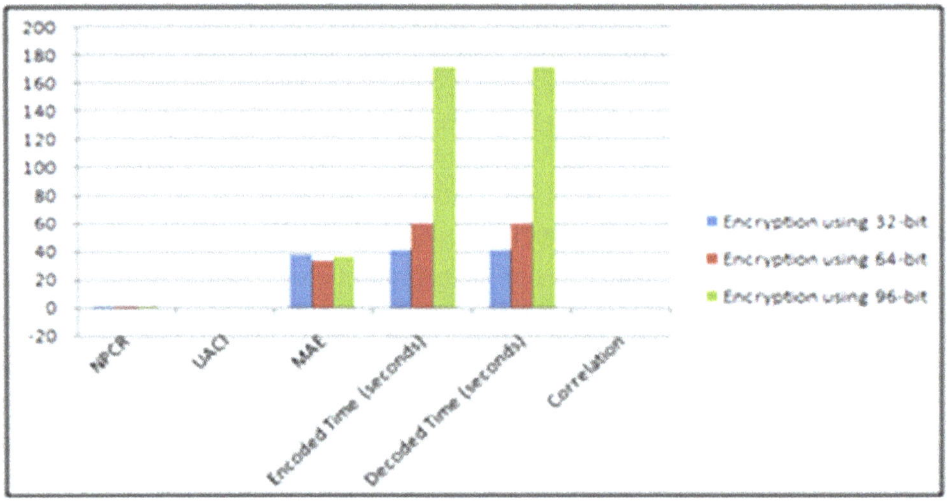

Figure 8.3(b). A graphical evaluation of the security parameters for the (256 × 256) size of test image.

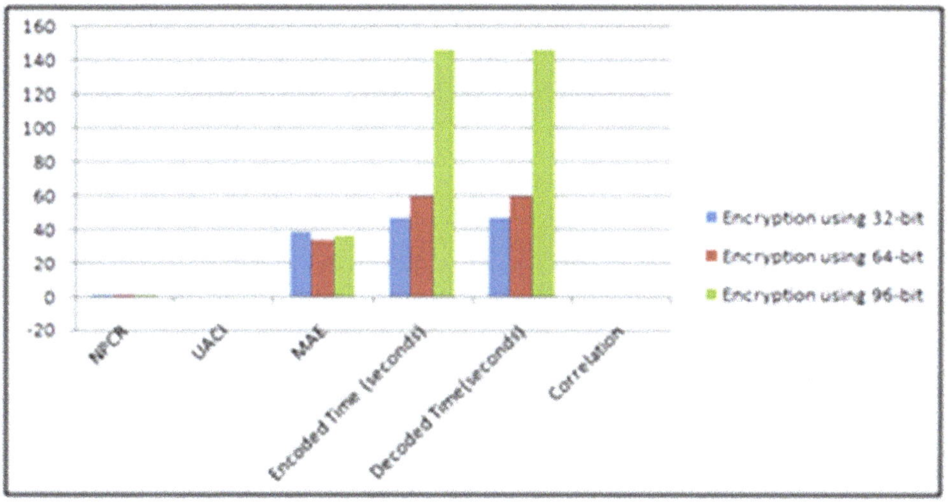

Figure 8.3(c). A graphical evaluation of the security parameters for the (512 × 512) size of test image.

Part B presents a quantitative evaluation of the security parameters, including the NPCR, UCAI, MAE, and the correlation coefficients of the proposed algorithm. A simulation is performed using images captured from the deep portion of the underwater environment (tables 8.1(a), 8.1(b) and 8.1(c)).

The experiment was performed using an Intel i3 processor with 2 GB of RAM and a 2 GB cache. Usually, captured images exhibit significant color-cast effects and inconsistent distributions of the RGB color channels due to irregular absorption of specific sections of the light spectrum by distributed sand particles. The encryption process of different test images using the proposed algorithm is as follows: the fundamental purpose of image encryption is to decrease the size for communication or storage space while at the same time maintaining appropriate features in the reconstructed images. High-quality encryption schemes have higher MAE values. Through the use of an encryption algorithm on an image, the pixels values alter as compared to the original image. A high-quality encryption algorithm should create such changes in an asymmetrical way in order to take advantage of the distinction in pixel values between the novel as well as the encrypted images. A good encrypted image, should be composed of completely random patterns so as obfuscate all the features of the original image. The encrypted images should be independent of each other and must have a low correlation with the originals.

Table 8.1(a). Performance analysis of the proposed hybrid cryptosystem using a 32-bit key.

Images	NPCR	UACI	MAE	Correlation
Image(1)	0.995	0.350 702	35.2078	0.095 6053
Image(2)	0.996	0.345 835	34.7191	0.096 4905
Image(3)	0.995	0.356 831	35.823	0.079 2723
Image(4)	0.996	0.331 194	33.2493	0.134 4840
Image(5)	0.995	0.325 613	32.689	0.128 2290

Table 8.1(b). Performance analysis of the proposed hybrid cryptosystem using a 64-bit key.

Images	NPCR	UACI	MAE	Correlation
Image(1)	0.995	0.391 381	39.2916	−0.035 8969
Image(2)	0.995	0.383 006	38.4507	−0.028 6331
Image(3)	0.995	0.394 198	39.5744	−0.047 8051
Image(4)	0.996	0.381 268	38.2763	−0.025 5326
Image(5)	0.995	0.377 919	37.9401	−0.056 9214

Table 8.1(c). Performance analysis of the proposed hybrid cryptosystem using a 96-bit key.

Images	NPCR	UACI	MAE	Correlation
Image(1)	0.995	0.378 751	38.0237	−0.000 7952
Image(2)	0.996	0.370 043	37.1494	−0.007 4844
Image(3)	0.996	0.382 426	38.3925	−0.024 5261
Image(4)	0.996	0.358 479	35.9885	0.011 4240
Image(5)	0.996	0.359 629	36.1039	−0.010 0232

Figures 8.3(a), 8.3(b), and 8.3(c) show comparative results for the security parameters of three different encrypted bit streams of the proposed hybrid underwater image cryptosystem using test images with sizes of 256 × 256 and 512 × 512 pixels, respectively.

8.6 Conclusions

In this chapter, we propose a hybrid underwater image cryptosystem which combines image enhancement (DHET) and the secure force algorithm (96-bit symmetric key generation). The proposed model employs a DHET for distorted underwater images, which restores visual quality, dynamically removes interference, and properly reconstructs the enhanced image. It combines a DCT with a tuned tri-threshold fuzzy intensification operator, which gives better contrast enhancement of underwater images. First, the background of the underwater image is detected by

DCT and the visual quality is improved using a tuned tri-threshold fuzzy intensification operator and Weber's law. Thus, the image contrast is enhanced and the interference is dynamically removed. Second, we employ an improved secure force algorithm based on a 96-bit symmetric key for underwater image security. This is a type of symmetric cryptographic algorithm which uses basic arithmetic operations to perform encryption and decryption. Hence, it has a Feistel structure to analyse the security inefficiency associated with the vehicle's image capture system. This mainly involves logical operations (AND, OR, XOR, XNOR), swapping processes, and the left-shift function. Thus, it causes minimum interoperability issues with the encoder and brings optimized image encryption and the benefits of secure transmission to operations deep underwater. In this chapter, we examined the efficiency of the proposed algorithm for encrypting images and compared it to existing standard algorithms. Performance evaluation metrics such as the PSNR, MSE, NPCR, UACI, MAE, encoding time, decoding time, and correlation coefficients were applied to both the existing and proposed algorithms.

Bibliography

[1] Farah M A B, Farah A and Farah T 2020 An image encryption scheme based on a new hybrid chaotic map and optimized substitution box *Nonlinear Dyn.* **99** 3041–64

[2] Preishuber M, Hutter T, Katzenbeisser S and Uhl A 2018 Depreciating motivation and empirical security analysis of chaos-based image and video encryption *IEEE Trans. Inf. Forensics Secur.* **13** 2137–50

[3] Zhang X, Seo S H and Wang C 2018 A lightweight encryption method for privacy protection in surveillance videos *IEEE Access* **6** 18074–87

[4] Eleftherakis D and Vicen R 2020 Sensors to increase the security of underwater communication cables: a review of underwater monitoring sensors *Sensors (Switzerland)* **20** 737

[5] Vreeland A, Kenneth R P, Henri (R) P, Bishop M, Kimberley M G, Matthew K D, Elliott S, Danielle M B and Chris B 2016 Considerations for exchanging and sharing medical images for improved collaboration and patient care: HIMSS-SIIM collaborative white paper *J. Digital Imag.* **29** 547–58

[6] Al-Husainy M A F and Al-Shargabi B 2020 Secure and Lightweight Encryption Model for IoT Surveillance Camera *Int. J. Adv. Trends Comp. Sci. Eng.* **9** 1840–47

[7] Sethi N and Vijay S 2013 Comparative image encryption method analysis using new transformed - mapped technique *Conf. on Advances in Communication and Control Systems* 2013 (Cac2s) pp 46–50

[8] Sowmyashree M S, Sukrita K B, Sneha R and Priyanka N 2014 A survey on the various underwater image enhancement techniques *Int. J. Engineering Sci. Invention* **3** 40–5

[9] Abouelseoud Y, Salama W M and Elkamchouchi H 2019 ArMTFr: a new permutation-based image encryption scheme *Int. J. Electron. Secur. Digit. Forensics* **11** 1–28

[10] Naik D K and Rout D K 2014 Outdoor image enhancement: increasing visibility under extreme Haze and lighting condition *Souvenir of 2014 IEEE Int. Advance Computing Conf. IACC 2014* (December 2015) pp 1081–6

[11] Politikos D V, Ioakeimidis C, Papatheodorou G and Tsiaras K 2017 Modeling the fate and distribution of floating litter particles in the Aegean Sea (E. Mediterranean) *Front. Marine Sci.* **4** 191

[12] Yin L, Chen B, Jinhua H and Ping C 2011 Research on the key technologies for underwater image communication system *Procedia Eng.* **24** 610–15
[13] Xie K, Wei P and Xu S 2018 An underwater image enhancement algorithm for environment recognition and robot navigation *Robotics* **7** 14
[14] Teng B and Zhao H 2020 Underwater target recognition methods based on the framework of deep learning: a survey *Int. J. Adv. Rob. Syst.* **17** 1–12
[15] Joseph J, Sivaraman J, Periyasamy R and Simi V R 2017 An objective method to identify optimum clip-limit and histogram specification of contrast limited adaptive histogram equalization for MR images *Biocybern. Biomed. Eng.* **37** 489–97
[16] Asamoah D, Ofori E, Opoku S and Danso J 2018 Measuring the performance of image contrast enhancement technique *Int. J. Comp. Appl.* **181** 6–13
[17] Huang H, Qirong T, Jiyong L, Wanli Z, Xuan B, Haitao Z and Gang W 2020 A review on underwater autonomous environmental perception and target grasp, the challenge of robotic organism capture *Ocean Eng.* **195** 106644
[18] Karimanzira D, Pfützenreuter T and Renkewitz H 2021 Deep learning for long and short range object detection in underwater environment *Int. J. Adv. Robot. Autom.* 1–10
[19] Bhadoriya D, Gupta R and Gupta M 2019 A block based scheme using tuned tri-threshold fuzzy intensification operators for underwater images *Int. J. Comp. Sci. Eng.* **7** 720–3
[20] Belmouhoub I, Djemaï M and Barbot J P 2003 Cryptography by discrete-time hyperchaotic systems *Proc. IEEE Conf. on Decision and Control* vol 2 pp 1902–7
[21] Guodong X H *et al* 2018 An image encryption algorithm based on time-delay and random insertion *Entropy* **20** 974
[22] Sharma A, Marriboyina V and Gupta K 2019 A novel underwater image enhancement technique *Int. J. Innov. Technol. Explor. Eng.* **8** 693–8
[23] Sharma A, Marriboyina V and Gupta K 2019 Design and implementation of key generation algorithm for secure image *Int. J. Recent Technol. Eng.* **8** 5139–46
[24] Qiao X, Bao J, Zhang H, Zeng L and Li D 2017 Underwater image quality enhancement of sea cucumbers based on improved histogram equalization and wavelet transform *Inform. Process. Agri.* **4** 206–13
[25] Khan S, Ibrahim M S, Khan K A and Ebrahim M 2015 Security analysis of secure force algorithm for wireless sensor networks arXiv:1509.00981
[26] Mali K, Chakraborty S and Roy M 2015 A study on statistical analysis and security evaluation parameters in image encryption *Int. J. Sci. Res. Develop. (IJSRD)* **3** 339–43
[27] Wu Y, Noonan J P and Agaian S 2011 NPCR and UACI randomness tests for image encryption *Cyber Journals: Multidisciplinary Journals in Science and Technology, Journal of Selected Areas in Telecommunications (JSAT)* **2011** 31–8
[28] Pallawi M, Motghare R and Agrawal P 2016 Performance analysis of encryption and decryption algorithms for securing images *Int. Res. J. Eng. Technol.* **03** 805–10

IOP Publishing

Advances in Image and Data Processing using VLSI Design, Volume 1

Smart vision systems

Sandeep Saini, Kusum Lata, Abhishek Sharma and G R Sinha

Chapter 9

A survey of thresholding in image processing

Eshaan Rathi and Abhishek Sharma

Thresholding is one of the simplest methods in image segmentation; it partitions or isolates image objects by converting grayscale images or colour images into binary images (images with only two pixel intensity values). It is an easy, yet effective way of partitioning an image into a foreground and background. It is based on assigning a threshold value to the pixel intensity of the original image. It is generally applied to grayscale images, but it can also be applied to color images. The following chapter provides background information and elucidates the need for thresholding in image processing. This chapter discusses various image thresholding methods, namely, histogram-based methods, entropy-based methods, clustering-based methods, local methods, and attribute-based methods. When required, this chapter documents the mathematical formulations of the techniques used.

9.1 Introduction

While dealing with digital image processing, we often come across the term 'image segmentation.' This is the process of dividing a digital image into multiple segments or groups of pixels and finally regrouping them. This is done so that analysis of the image becomes easier as we convert it into a meaningful format. It is also done in order to find features, patterns, objects, or boundaries in an image. Every pixel in an image has a specific value; in segmentation, we assign that value in such a way that all pixels with the same value lie in the same group; this means that all pixels in the group have some common characteristics and attributes, such as colour, intensity etc.

Generating an algorithm for precise image segmentation is important, but it can also be very challenging. One way to segment images is via thresholding. As discussed earlier, it is the foremost and easiest way to do so, and is applied to grayscale images. More formally, a threshold value, T, is chosen. If the intensity of a

doi:10.1088/978-0-7503-3919-3ch9

pixel $I_{x,y}$ is <= T, the pixel is replaced by zero (black), otherwise, it is replaced by one (white). Figure 9.1 illustrates the thresholding process for a random image with pixel intensities from 0 to 255 subjected to thresholding with a threshold value of 100.

As per Sezgin and Sankur [1], thresholding methods are classified based on the data retrieved and obtained from the information, namely, histogram-based methods, entropy-based methods, clustering-based methods, attribute similarity methods, and local methods. We discuss these methods in the following sections.

9.2 Histogram-based methods

The challenge for any thresholding process is the process used to choose a threshold value. The threshold value can be manually chosen, an algorithm can be used to compute the value, or a mean or median value can be used.

Histogram-based methods create a histogram of the image pixel intensity; the shape of the histogram is used to determine the threshold value. Generally, the peaks of the smoothed histogram are used to arrive at the threshold value, but troughs and bends can also be analyzed and used in this method.

This method of image sementation is mostly used when there are mainly uniform intensity regions in the image being analysed. In this case, the threshold is used to separate the background from the region of interest. The entire image is scanned pixel by pixel and its grey value is checked with respect to a minimum value which is called the threshold.

Rosenfeld's method [2] is based on comparing and analyzing the curves of the histogram h(g), i.e. the concave versus the convex. When the convex hull of the histogram has been computed, the minima points can be considered as possible threshold values. Sezan, in his paper [3], performs a peak analysis by superposing and combining h(g) with a smoothing and difference kernel. Furthermore, after superpositioning, the peaks are combined and smoothed, reducing the histogram to a two-lobe function. The grey levels are estimated at the points where the peaks begin and where they attain their maxima. To optimise the data, the algorithm

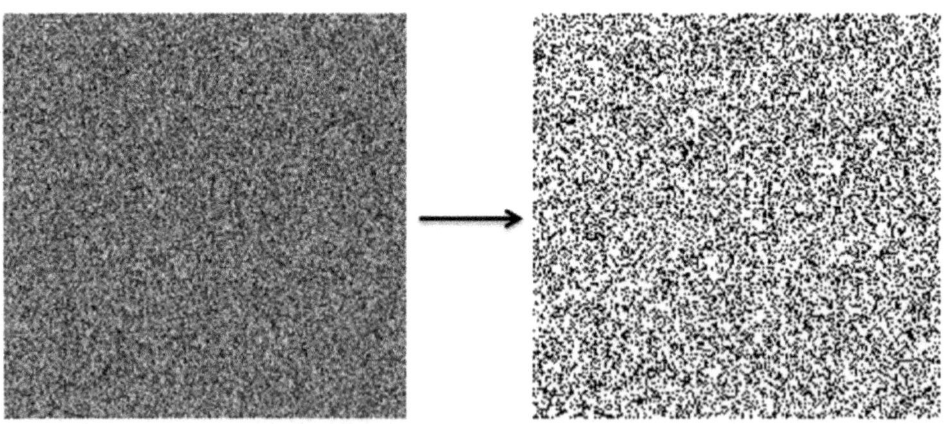

Figure 9.1. An example of thresholding.

determines the grey value threshold between the apertures and the quantization level is chosen such that the grey levels peak and attain their maximum values.

Ramesh *et al* [4] used a simple functional two-step approximation to the histogram. In this approximation, an iterative search process is used to arrive at the threshold by minimizing the sum of the squares between the histogram and the function.

Cai and Liu [5] have presented an approximation to the spectrum as the power spectrum of complex exponential signals.

9.2.1 Balanced-histogram-based thresholding

This method is also based on histogram techniques, but in this case, the image is partitioned into two parts, i.e. the foreground and the background, and the optimum threshold values are then found by processing and balancing these two parts.

First, imagine that we have a histogram and we are weighing it on a scale (i.e. measuring the amounts of foreground and background). We get a heavier side and a lighter side. We remove some weight from the heavier side and then try to balance it again. This process continues until the histogram is balanced between the foreground and the background.

9.3 Clustering-based methods

This method works on the principle of generating clusters from the grey-level data. The clusters which are generated from the grey levels are then analysed. Since the final result is a binary image, the number of clusters is always two. Each cluster represents a different lobe of the histogram. Clustering-based methods are further divided on the basis of their approaches.

9.3.1 Iterative clustering

Iterative clustering is based on a combination of two Gaussian-based models. The steps involved in the algorithm are as follows:

1. At every iteration, a new threshold (T) is arbitrarily chosen.
2. The image is segmented into a foreground (FG) and a background (BG) on the basis of T.
3. The mean of the FG set is found (t_1), the mean of the BG set is found (t_2), and then the mean of t_1 and t_2 is found (t_{avg}).
4. Steps 1–3 are repeated until the difference between T and t_{avg} reduces and becomes extremely small.

A modification of this algorithm is to set the initial value of the threshold to the mean of the highest grey level and the lowest grey level of the histogram (i.e. the mean of the two peaks of the histogram).

9.3.2 Clustering thresholding

This method of clustering was first described by Otsu [6] and is one of the most referenced and commonly used thresholding techniques, due to the ease of its computation process. The Otsu algorithm gives a uniform intensity threshold that separates the image into two categories of pixel, the foreground and background pixel classes. The optimum value of the threshold is found by minimizing the variance intensity within each class (the intra-class variance), which is similar or equivalent to maximizing that between classes (the inter-class variance).

The intra-class variance can be defined as the weighted sum of the variances of the foreground and the background:

$$\sigma^2_{\text{intra}} = p_{\text{F}}(T)\sigma^2_{\text{F}}(T) + p_{\text{B}}(T)\sigma^2_{\text{B}}(T)$$

where p_{F} and p_{B} are the probabilities of the foreground and background classes separated by a threshold value, t, and σ^2_{F} and σ^2_{B} are the respective variances.

This requires the computation of intra-class variance to obtain the threshold values of each class. This can be an expensive process, due to the large amount of computation required; hence, this might not be a viable option. To overcome this issue, calculation of the inter-class variance can be employed, which is a comparatively economic process. It is calculated by subtracting the intra-class variance from the total variance.

$$\sigma^2_{\text{inter}} = \sigma^2 - \sigma^2_{\text{intra}} = p_{\text{F}}(T)p_{\text{B}}(T)[u_{\text{F}}(t) - u_{\text{B}}(t)].$$

The maximum of σ^2_{inter} is the desired threshold.

To find the threshold, the algorithm computes the histogram and probabilities for every intensity level. It then initializes the probabilities and means for $t = 0$. It then iterates over each possible threshold from 0 to the maximum intensity value and updates the probabilities and means for each t and computes the corresponding variances. The maximum variance is the required threshold.

One of the drawbacks of Otsu thresholding is that significant details may be lost, which can be of prime importance in further processing.

9.3.3 Minimum-error thresholding

This method minimizes the error in classification by assuming that every cluster or group has a Gaussian distribution with a mean and a standard deviation that are independent of the threshold chosen.

Unlike Otsu's method, which computes the mean and the variance by segmenting the image into two clusters based on the threshold values, this method assumes a normal distribution and estimates its parameters based on this assumption. If the two Gaussian distributions are separated or even have some overlap, it assumes that if an arbitrary threshold is chosen, the mean and variance of every cluster are good approximations to the mean and variance of the original or fundamental populations. The optimal or ideal threshold is chosen by observing the threshold for which the mixture of the two Gaussians is the best approximation of the actual histogram.

To minimise the complexity of this method, a gradient descent method can be used to make the initial estimate. Methods such as Otsu's are a good way to make an initial estimation.

9.3.4 Fuzzy clustering thresholding

This method approaches the thresholding task by allocating fuzzy clusters to pixels in the image, based on the difference of the mean of two classes. Ramesh [4] proposed two thresholding methods. The first minimizes the variance of the approximated histogram, whereas the other minimizes the sum of the squared errors. The first method turns out to give better results as compared to the second, although it adds a little complexity.

9.3.5 *k*-means clustering thresholding

The primary idea of this method is to reduce the sum of the distances between the data points and their corresponding clusters. In this technique, the value of k is predetermined and chosen by the algorithm. An unlabeled data set is then taken as an input by the algorithm, and this data set is then divided into clusters, called the k clusters. This process is repeated until it stops finding the best data cluster.

The steps involved in this algorithm are as follows:

1. A value of k is predetermined.
2. Initial clusters are arbitrarily chosen.
3. Each data pixel is assigned to the closest cluster center.
4. For each data point obtained, redetermine and recalibrate the centers of the clusters. This is also needed for clusters which lose data points.
5. Repeat steps 2 to 4 until a convergence of distance criteria is obtained.

When the mean of the kth and the $(k-1)$th class centers is determined, this value is the threshold value.

9.4 Entropy-based thresholding

The idea of entropy-based thresholding originates from the second law of thermodynamics. This method uses the entropy of the distribution of grey levels in an image. This is a measure of the spontaneous dissipation of energy. A large entropy value is an indication of a large information transfer, while a low entropy value means less transfer of information, or the conservation of information. This is obtained by reducing the cross-entropy between the grey-level input and the digital image output.

Kapur *et al* [7] employed entropic concepts taken from digital communications and signal processing, which were originally described by Shannon. In this method, the entropies corresponding to the background and foreground image signals are computed and summed. The threshold is considered to be at its optimum when the sum of these values reaches a maximum.

The probability distribution of the grey levels of the image with respect to the black region is:

$$P_0/P_{\text{Black}},\ P_1/P_{\text{Black}},\ \ldots,\ P_t/P_{\text{Black}},$$

and similarly, that with respect to the white region gives:

$$(P_t+1)/1-P_{\text{Black}},\ (P_t+2)/(1-P_{\text{Black}}),\ \ldots,\ (P_n-1)\ /\ (1-P_{\text{Black}}).$$

Here, t represents the threshold and P_{Black} represents the probability of the grey level, which is less than or equal to t, while P_i represents grey-level i's pixel probability.

$$P_{\text{Black}} = \Sigma^{s}{}_{i=0}\, p_i.$$

The entropy of the image is $H_{\text{Black}} = -\Sigma^{s}{}_{i=0} pi/P_{\text{Black}} \log(p_i/P_B)$.
The entropy of the background is

$$H_{\text{white}} = -\left(\Sigma^{n-1}{}_{i=s+1} pi/\ (1-P_{\text{Black}})\log\left(p_i/(1-P_{\text{Black}})\right)\right).$$

The threshold s is defined as the maximum of the sum of $H_{\text{Black}} + H_{\text{White}}$. This sum essentially represents the total entropy. It has been experimentally found that this value is at a minimum when the difference between the reconstructed and the observed images have similar mean intensities in both their foregrounds as well as their backgrounds.

There are various types of entropy-based thresholding, such as cross-entropy thresholding and fuzzy-entropy thresholding.

9.4.1 Cross-entropy thresholding

This method determines the threshold value by minimizing the information's theoretical distance. This is basically obtained by decreasing the cross-entropy between the output binary image and the input grey image. It can also be obtained by increasing and maximizing the cross-entropy between the input and the output image.

9.4.2 Fuzzy entropy thresholding

This method uses fuzzy memberships as a criteria for a grey value to be part of either the background or the foreground. The more distant a gray value is from an assumed threshold, the greater its possibility of belonging to a specific class.

9.5 Attribute similarity methods

In these methods, the threshold value is decided by the similarity and parity of the original image with that of the digital binary image, based on some characteristic, attribute, or quality. For example, these attributes could be the shape, the matching of the edges of the objects, the stability of the objects, grey levels, texture, segmentation, etc.

9.6 Local method

In the local method, a different and distinctive threshold number is computed for each and every pixel. This is more or less decided by the grey scale information of the adjacent pixels. The threshold value is computed using some local parameters, such as deviation or range. The local parameters are used to segment the image into several parts. The threshold value is computed for these parts using bimodal histograms.

The threshold values of each part are analysed, and, using these, the threshold is estimated for the whole image. This method gives more accurate results than a fixed threshold value applied to the whole image. This approach has been used in TV images of machine components. One of the disadvantages of the local method is that it is time-consuming; hence, a hybrid approach can be used.

9.7 Conclusions

Image thresholding is a vast topic and one of immense importance. It has numerous applications in multifarious areas, such as image processing, video processing, medical imaging, augmented reality, engineering design, AI-based surgery, face identification, etc. This chapter discussed the importance of image thresholding and elaborated various image thresholding techniques categorized by their approaches.

Bibliography

[1] Sezgin M and Sankur B 2004 Survey over image thresholding techniques and quantitative performance evaluation *J. Electron. Imaging* **13** 146–68

[2] Rosenfeld A and De La Torre P 1983 Histogram concavity analysis as an aid in threshold selection *IEEE Trans. Syst. Man Cybern.* **SMC-13** 231–5

[3] Ibrahim Sezan M 1990 A peak detection algorithm and its application to histogram-based image data reduction *Comput. Vision Graph. Image Process.* **49** 36–51

[4] Ramesh N, Yoo J and Sethi I 1995 Thresholding based on histogram approximation *IEE Proc. Vision, Image and Signal Processing* **142** 271–9

[5] Jinhai Cai and Liu Z-Q 1998 A new thresholding algorithm based on all-pole model *Proc. Fourteenth Int. Conf. on Pattern Recognition (Cat. No. 98EX170)* vol 1 pp 34–6

[6] Otsu N A threshold selection method from gray-level histograms *IEEE Trans. Syst. Man Cybern.* **9** 62–6

[7] Kapur J N, Sahoo P K and Wong A K C 1985 A new method for gray-level picture thresholding using the entropy of the histogram *Comput. Vis. Graph. Image Process.* **29** 273–85

IOP Publishing

Advances in Image and Data Processing using VLSI Design, Volume 1

Smart vision systems

Sandeep Saini, Kusum Lata, Abhishek Sharma and G R Sinha

Chapter 10

Review of quality assessment of fruit and vegetables using NIR spectroscopy

R Mariappan, B Lakshmi Sirisha and P S Suhasini

Fresh fruit and vegetables contain minerals and vitamins, and their percentages of fat, salt, and sugar are low. It is very important to eat them every day for good health. They are also a very good source of dietary fiber. A high intake of fruit and vegetables helps to reduce cholesterol,[1] blood pressure, heart disease, cancer, etc. Fresh fruit and vegetables increase the immunity of the human body, helping it to fight diseases such as COVID-19. So a continuous supply of undamaged, fully ripe, and nutritious fruit is required to satisfy the demands of the increasing population. Quality checking and the grading of fruit and vegetables are essential at different stages of processing. In past, inspection was performed by manual and destructive methods, which were preferred for assessment. In the digital world, the assessment is performed using technology. Most of the food-processing industry uses various optical methods, electromagnetic methods, and non-destructive methods. These techniques have the benefits of simplicity and speed. This chapter elaborates a non-destructive method based on the principles of near-infrared (NIR) spectroscopic techniques. The different spectral modes used for the assessment of fruit, i.e. reflectance, interactance, and transmittance, are explained. We address the measurement of different attributes of multiple kinds of fruit and vegetables using optimal variable selection and calibration methods. The main objective of this review is to present the role and importance of NIR spectroscopy in the assessment of various fruit quality attributes.

10.1 Introduction

India has secured second place in the world for the production of fruit and vegetables, and has first place for the production of a few fruits and vegetables,

[1] https://www.betterhealth.vic.gov.au/health/conditionsandtreatments/cholesterol

namely banana, papaya, mango, cashew nuts, and potato. According to the National Horticulture Board, India supplied 169.1 million metric tonnes of vegetables and 90.2 million metric tonnes of fruit in the year 2015–2016. However, India contributes only 1% out of 90% to the global market due to poor harvest management techniques [1]. Meanwhile, wastage of approximately 30 to 35% of fresh fruit is observed in the various stages of food processing.

During the COVID-19 pandemic, people have been extremely aware of the importance of healthy food. Healthy foods have numerous healthy constituents, such as vitamins, proteins, minerals, etc. Moreover, consumers also attach importance to certain quality attributes, such as freshness, exact ripening, total soluble solids, acidity content, etc.

In the past, the quality of these products was tested by manual inspection, and destructive methods were used to analyze the quality. Today, to reduce the problems caused by manual and destructive methods, some spectroscopic techniques have been successfully launched in the food industry.

The sensory method is the oldest method, but it is still used every day. This analysis is performed by both consumers and food observers. It requires a large number of samples; its results are typically subjective and difficult to measure quantitatively. The use of chemicals is a destructive method; it is not suitable for fast, online product management. This method only uses a sample, so if any sample is destroyed, that result reflects on the remaining samples as well. Hence, this method does not produce accurate results. Microbiological techniques are mainly used to detect fungi and bacteria, but not for identifying other contents; they are suitable for slow and offline measurements only. Moreover, these processes are complex and more expensive than other methods. To overcome all these limitations, fast and non-destructive optical methods were introduced. NIR spectroscopy is also the best optical method to estimate the properties of fruit and vegetables accurately. However, spectral measurement is not a easy process; it requires expertise. A comparison of all these methods is shown in table 10.1.

10.2 NIR spectroscopy

NIR spectroscopy was first used in agriculture by Norris in 1964 to measure the moisture content of grain [1]. It was subsequently used to measure the moisture, fat, and protein contents of different agricultural food products [2, 3]. The spectrum of NIR radiation falls in the range of 780–2500 nm. The benefits of NIR spectroscopy

Table 10.1. Comparison of the methods used to detect fruit quality.

Name of technique	Type of process	Amount of sample
Sensory technique	Subjective process	Entire product
Chemical technique	Objective process	Sample of the product
Microbiological technique	Objective process	Sample of the product
NIR technique	Objective process	Entire product

techniques are: (1) they are non-destructive methods—the internal quality parameters of the product are tested without damaging its surface, (2) the measurement is completed very easily and rapidly, (3) no chemicals are used on the fruit samples, (4) it is possible to simultaneously measure the internal attributes of several fruits. One of the limitations of spectroscopy is that it cannot provide spatial information.

10.2.1 NIR spectrometer

A schematic of a conventional NIR spectrometer [4] is shown in figure 10.1. **Light source:** in NIR spectroscopy, tungsten lamps are used as a light source; these are low-cost and produce high-intensity continuous wavelengths. In some cases, LEDs are used instead of tungsten lights. **Beam-splitter system:** this system plays a major role in NIR spectrometers; it translates multiple wavelengths into single-colored light (it acts as a light filter).

Sample detector: the sample detector varies considerably according to the type of sample used. For liquid samples, a glass or quartz chamber is used; for solid samples, a diffusive–reflective carrier is used. In real-time applications, a fiber-optic cable is preferred, as shown in figure 10.2.

A computer with software is used for data acquisition, analysis control communications, and numerical analysis in the spectrometer system.

10.2.2 Modes of data acquisition in NIR spectroscopy systems

There are three modes of data acquisition, namely reflectance, transmittance, and interactance.

Reflectance mode: in this mode, the light source and the detector are placed on the same side of the sample so as to capture the reflected light from the sample, as shown in figure 10.3. This mode is used for solid or granular samples. In the reflectance mode, the sample fruit is illuminated by a laser light powered by a DC regulated power supply. The spectrum is detected by the spectrometer. In order to avoid specular reflection in the reflectance mode, the light source and the detector are placed at an angle of 45° to the sample.

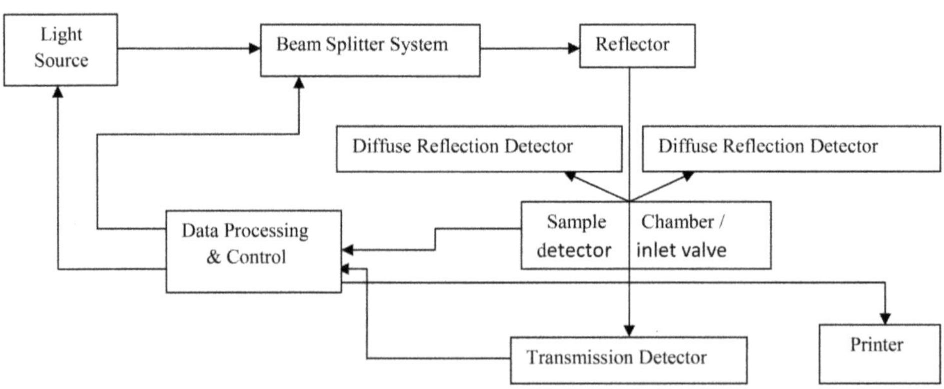

Figure 10.1. NIR spectrometer schematic.

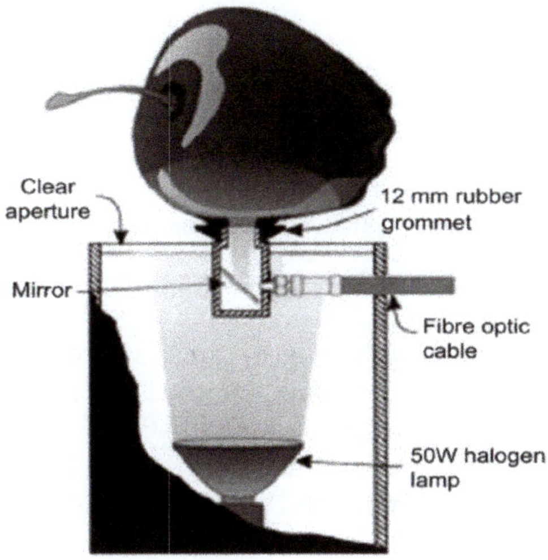

Figure 10.2. Visible NIR measurement system.

Figure 10.3. Reflectance mode.

Figure 10.4. Transmittance mode.

Transmittance mode: in this mode, the detector and illumination source are placed on opposite sides of the sample. This mode is generally used for liquid sample analysis and also for some solid samples, such as grains and dairy foods, as shown in figure 10.4.

Interactance mode: interactance mode, on the other hand, is a combination of reflectance and transmittance. In this mode, the light source and the detector are situated parallel to each other, as shown in figure 10.5. Large samples, such as whole fruits, can be tested in this mode. In addition, as a result of research, it has been observed that, compared to the reflectance and interactance modes, the transmittance mode yields better accuracies for classification and also for internal disorder detection.

Figure 10.5. Interactance mode.

In general, the reflectance wavelength range is from 450 to 1100 nm, the interactance wavelength range is from 520 to 1100 nm, and the transmission wavelength range is from 690 to 1080 nm [4].

NIR measurements [5, 6] are combined to form data sets, such as reflection measurements, transient measurements, and transmission measurements at 90° (T90) and 180° (T180). Chemometric methods are applied to NIR measurements in order to extract information related to the quality attributes and also to eliminate the interference of factors which are not related to the sample concentration. In general, chemometrics [7, 8] can be divided into two types: spectral pretreatments and regression methods.

10.3 Chemometrics

Chemometrics are used to extract the chemical and physical information from complex data by mathematical and statistical techniques. Water is the major constituent of fruit and vegetables. Fruit quality attributes are measured using the NIR spectrum; however, infrared radiation is affected by water. Furthermore, the NIR spectrum has a low peak signal-to-noise ratio (PSNR). All of these factors cause convolution of the NIR spectrum. Therefore, chemometrics are applied to remove the noise from the spectrum and measure the attributes of the fruit and vegetables.

10.3.1 Preprocessing methods

10.3.1.1 Smoothing

Smoothing [6, 9] is used to remove the high-frequency noise from a spectrum, thereby improving the signal-to-noise ratio (SNR). As per the observations of Sun *et al* [10], smoothing is the most reliable preprocessing method for detecting the soluble solid content (SSC) in oranges. The spectrums are acquired and averaged, as averaged spectrums reduce the thermal noise produced by the detector.

10.3.1.2 Offset correction

Instrument noise is reduced using this method. It adjusts the baseline drift, leaving the spectrum shape unchanged. In each spectrum, the first few wavelength points are averaged and then corrected using this correction.

10.3.1.3 Derivative correction
The first derivative is used to eliminate drift and the second derivative is used to eliminate scattering effects. These derivatives are very useful for removing background interference. They also improve the spectral sensitivity and resolution. As per the analysis of Pissard *et al* [11], Savitzky-Golay (SG) first-derivative processing yields good results as a preprocessing method. As per the analysis of Liu *et al* [12], second-derivative processing yields good results.

10.3.1.4 Normalization
Multiple scatter correction [13, 14] is popular technique for normalization. This technique compensates for both additive and multiplicative effects in the spectral data. Extended multiple scatter correction compensates for the chemical interference effects obtained by incorporating the known spectra of the interference.

10.3.1.5 Orthogonal signal correlation
Some principal components (PCs) obtained by the partial least-squares regression (PLS) and principal component analysis (PCA) methods [15–18] have very little information related to the concentration matrix. There is no use for such principal components; hence, they are removed using various orthogonal approaches, which enhances the predictive ability and stability of the calibration model. Orthogonal signal correlation is one of the best models for improving the predictive ability.

10.3.1.6 Transformation technique
Chemical signals have various frequencies; based on this fact, these signals can be decomposed into different scale compositions. This decomposition can be performed effectively using a wavelet transform. Narrow and wide windows are used to observe drastic changes and the overall features of the spectrum, respectively.

10.4 Calibration methods

10.4.1 Multiple linear regression (MLR)

This type of regression method [19, 20] estimates the variables at every wavelength using spectral values in a linear combination. This is a statistical method, which uses several variables to predict the outcome of a response variable. Regression values are predicted in a least-squares context by reducing the prediction error.

10.4.2 Principal component regression (PCR)

PCR is mainly used to estimate the coefficients of unknown regressions. An analysis of the principal components is used to estimate these coefficients. These components are applied as predictors, instead of the original spectral data, and used to fit an MLR model.

10.4.3 Partial least-squares regression (PLS)

The PCR method has several advantages; however, its main drawback is that the decision about how many principal components to keep does not depend on the response variable. To overcome this drawback, PLS was developed by Wold *et al* [7]. PLS estimates the dependent variables by extracting the smallest possible set of orthogonal factors.

10.4.4 Least-squares support vector machine (LS–SVM)

LS–SVM denotes the least-squares version of SVM. This is a supervised learning method that analyzes data and recognizes patterns; it is used for classification and regression analysis.

10.4.5 Artificial neural network (ANN)

ANNs are used in NIR calibration. They consist of three layers of neurons: the input, output, and hidden layers [3]. In this method, every neuron in a given layer is connected to every neuron in the next layer. Every connecting line has a weight factor whose value is assessed based on a calibration set using cross-validation. ANN achieved good results in the studies of Monika Jhuria *et al* [21] and Pushpa *et al* [22]. However, ANN also faces some drawbacks; for example, it takes a lot of time to train the data, it faces visualization difficulties, and it suffers from overfitting problems.

A survey of the literature on the topic of fruit quality detection, based on international publications, is shown in table 10.2; a survey based on national publications is shown in table 10.3. These surveys demonstrate that NIR spectroscopy accurately measures the quality attributes of fruit and vegetables, compared to traditional methods.

10.5 DLP technology in near-infrared (NIR) spectroscopy

Digital light processing (DLP)-based spectrometers measure the spectrum of light incident on an input port. In this process, the light from an input slit is dispersed to separate the wavelengths and then reimaged on a digital micromirror device (DMD) (Ref: Application Report—Texas Instruments) [45, 46]. To measure the spectrum distributed across the DMD, the processing steps include pattern generation, detector sampling, signal averaging, and spectrum decoding. Patterns are generated by the desired spectral filter for application to the light at the system's input aperture. The patterns are then loaded onto the DLP controller and sequentially displayed on the DMD. The samples are subsequently averaged to give a single value for each DMD pattern. The averaged detector values are then processed to get a spectrum.

The advantages of NIR spectroscopy with DLP technology are as follows:

- the performance is high
- the cost is low, as single-element detectors and low-cost optics are used
- the captured signal is very strong
- the light intensity incident on the detector may be varied

Table 10.2. A survey of the literature on the topic of fruit quality detection, based on international publications.

S. no.	Authors	Proposed method	Methodology	Remarks
1	Amodio *et al* [23]	Fourier transform (FT)-NIR spectrometer, partial least-squares discriminant analysis (PLS-DA) model	Manual, rotating the fruit in all directions. The three scanners scan the spectrum originating from the fruit; the average of these spectra was considered.	Only for strawberries
2	Carames *et al* [24]	NIR spectroscopy, PLS	To minimize the light scattering effect, they used multiplicative scatter correlation. The spectrum data were processed by the PLS method.	This method decreases the time and wastage.
3	Pissard *et al* [25]	NIR spectroscopy and reference methods	Each apple is subjected to NIR spectroscopy; the spectrum in measured in different directions, and the averaged spectrum is considered.	Only for apples
4	Yu *et al* [26]	PLS regression	The light beam falls on grapes, which reflect some energy. That reflected energy is captured and used in spectroscopy.	Only for grapes
5	Jaiswal *et al* [19]	PLS and MLR	First-, second-, and third-order derivatives and smoothing of an MLR regression model were developed and tested using cross-validation.	Only for bananas
6	Rajkumar *et al* [20]	Hyper-spectral imaging technique in the visible and near-infrared (400–1000 nm) regions and MLR	In this paper, the quality attributes were predicted using MLR. The validity of the observed data is tested using PC analysis.	Only for bananas
7	Subedi *et al* [27]	Visible–short wavelengths, near-infrared spectra	All stages of maturity of mango fruit are modeled by DM.	This method is recommended for assessment of the ripeness of mangoes.

8	Guthrie *et al* [28]	NIR spectrophotometer	A fiber-optic light is focused on the fruit, which then reflects some energy that is collected by an NIR spectrometer.	Only for mangoes
9	Mendoza *et al* [29]	NIR spectroscopy, PLS method	In this paper, scattering profile analysis is performed by wavelet transforms.	Only for apples
10	Zude [30]	Near-visible (VIS) and NIR in reflectance mode	Measurement of the spectral reflectance, wet-chemical analysis, calibration, and statistical analysis.	Only for bananas

Table 10.3. A survey of the literature on the topic of fruit quality detection using NIRS, based on national publications.

S. no.	Authors	Proposed method	Methodology	Remarks
1	Ansari *et al* [31]	Bio-speckle techniques	A laser beam was focused on the fruit; bio-speckle patterns were collected using a digital camera.	This method is not suitable for all fruits. It is suitable for apples.
2	Sindhi *et al* [32]	Extraction of features and classification	Feature vectors were generated based on attributes such as color, texture, and shape in the fruit image. Then the fruit was classified.	Diseases found in apples only
3	Jadhav *et al* [33]	A fruit size detection and grading system based on image processing	Find the center coordinate of the fruit's shape in the image. Find the fruit's axis in the image	Takes more time for processing.
4	Bhargava *et al* [34]	Segmentation, feature extraction, and classification	Non-destructive methods are used to assess the quality of food.	Wastage is high.
5	Jhuria *et al* [21]	ANN, back-propagation algorithm used for classification	Identify diseases on fruits during formation.	ANN was a successful grading technique for apples and mangoes.
6	Bhange *et al* [35]	*k*-means algorithm and SVM	Classify the affected fruits using a clustering algorithm.	Effective for pomegranate disease only

7	Pushpa *et al* [22]	*k*-means clustering technique, blob extraction method, template matching method.	Features extracted from segmented regions of the images are used for training and stored in a feature database; classification then is performed.	Only considers apple diseases.
8	Dubey *et al* [36]	*k*-means clustering technique, multiclass SVM.	An image processing approach is used for fruit disease identification.	Classifies fruit with good accuracy.
9	Pandey *et al* [37]	Machine learning and deep learning techniques.	Quality assessment is performed using the size and type of fruit using image processing techniques.	Considers pomegranate fruit.
10	Payel *et al* [38]	Soft x-ray emission spectroscopy	The internal structure of pomegranate fruit is identified using x-rays.	Internal defects in pomegranate fruit only.

- light is captured at multiple wavelengths
- selection of the specific wavelengths of interest is performed using custom spectral filters.

10.5.1 Proposed fruit and vegetable quality detection system

The proposed fruit and vegetable quality detection system, which uses a DLP-based NIR spectrometer, is shown in figure 10.6.

The subsystems in this quality detection system are (i) a DLP-based spectrometer, (ii) spectral data preprocessing, (iii) a calibration model, and (iv) model validation

10.5.1.1 Operation of the spectrometer

In reflectance mode, when broadband light is directed onto a sample, some light is absorbed by the sample and some light is reflected. The light observed at each

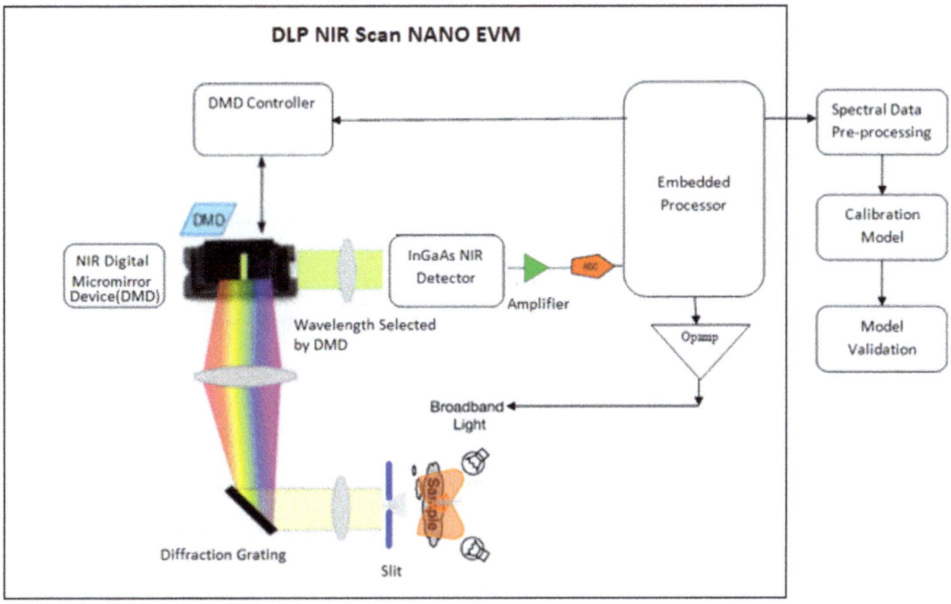

Figure 10.6. A fruit and vegetable quality detection system which uses a DLP-based NIR spectrometer.

wavelength, which depends on the molecular structure of the sample, is called the chemical fingerprint. The light reflected from the sample is collected at the lens and focused on the input slit. The size of the slit is selected to balance the wavelength resolution and the SNR of the spectrometer. The light passing through the slit is collimated by lenses, and the passes through a wave pass filter. When it hits a reflective grating, is it dispersed into its constituent wavelengths. The averaged detector values are then processed to get a spectrum. After the spectrum is obtained, chemometric methods are applied in order to get information about the quality attributes. Unrelated factors are eliminated.

10.5.1.2 Spectral data preprocessing

During the equalization of the spectrum, noise and background data are also obtained along with the useful data. A preprocessing stage is required to remove the unwanted data and obtain reliable and accurate information.

10.5.1.3 Calibration models for quantitative and qualitative analysis

Calibration models are very much required to estimate the quality of fruit. Regression techniques are used to predict the selected maturity indices of fruit and vegetables. PLS regression, MLR, PCR, and artificial neural networks (ANNs) are the techniques [2] that will be investigated to select the method most appropriate for fruit and vegetables. Supervised approaches, LS–SVM, linear discriminant analysis (LDA), and partial least-squares discriminant analysis (PLS-DA) are the other methods used.

10.5.1.4 Model validation

The twofold cross-validation method partitions the samples equally into training and test samples. The particle swarm optimization (PSO) algorithm is used to optimize the accuracy of the SVM classifier.

10.5.1.5 Model evaluation

The calibration accuracy and validation values are measured using the following metrics:

$$\text{Correlation coefficient r} = \frac{\sqrt{\sum_{i=1}^{n} (\widehat{y}i - yi)^2}}{\sqrt{\sum_{i=1}^{m} (\widehat{y}i - ym)^2}} \tag{10.1}$$

where y_i is the measured value and $\hat{y}_i$ is the predicted value of the ith sample; y_m is the mean value.

Here, n is the number of observations in the data set; it equals

$$\sqrt{\frac{1}{n_c}\sum_{i=1}^{n_c} (\widehat{y}_i - y_i)^2} \tag{10.2}$$

where n_c is the calibration set. N_p is the prediction set; it equals

$$\sqrt{\frac{1}{n_p}\sum_{i=1}^{n_p} (\widehat{y}_i - y_i)^2}. \tag{10.3}$$

$$\text{Ratio performance deviation(RPD)} = \frac{\text{SD}}{\text{RMSEP}}. \tag{10.4}$$

SD is the standard deviation of the validation data set.

10.6 Fruit and vegetable quality parameters

The following are the parameters to be measured when estimating the quality of fruit and vegetables.

Both external and internal quality parameters, such as colour, texture, bruise damage, etc. can be measured using NIR spectroscopy.

10.6.1 Soluble solid content (SSC)

Along with sugar, fruits have acids, vitamins, proteins, and minerals. These are all known as soluble solid contents.

10.6.2 Sugar content

All fruit has a sugar content. Sweetness is the one of the attributes of ripe fruit.

10.6.3 Firmness

Firmness is a quality parameter in grading fresh produce, which is governed by the mechanical and structural properties of fruit.

10.6.4 Dry matter

Fresh fruit is rich in water; if fruit, vegetables, or roots have no water content then they are known as dry matter (DM). Small values of DM mean that the fruits or vegetables contain good amounts of sugar, acids, vitamins, proteins, and minerals.

10.6.5 Titratable acidity

Titratable acidity measures the total acid concentration in fresh produce and is determined by titration of intrinsic acids with a standard base.

10.6.6 Carotenoid content

Carotenoids are plant pigments responsible for bright red, yellow, and orange hues in many fruits and vegetables. Lycopene is a plant nutrient with antioxidant properties; it is a pigment responsible for the characteristic colour in vegetables and fruits such as tomatoes, grapes, watermelons etc., On the other hand β-carotene is a red-orange pigment.

Furthermore, this method can also detect insects, which cause damage to fruit and vegetables. NIR transmission spectra can identify wax components on the surface of fruits; waxing is a very popular storage technique that is particularly used for storing apples.

10.7 Discussion

According to the overviews above, most researchers focused on spectroscopic techniques to evaluate fruit quality. Spectroscopy provides good accuracy and a robust calibration model compared to traditional methods. Here, a large database is a very important requirement; it should include different weather conditions, different origins, etc. to yield good, stable, and accurate results. Research work has mostly been done using the major attributes mentioned in section 10.6. It is observed that the main disadvantage of using NIR spectroscopy for fruit quality assessment is the low precision of its reference methods. In most cases, NIR spectroscopy needs more accurate calibration than a reference method.

In the international publications shown in table 10.2, most of the research involved fruits such as strawberries, kiwifruit, grapes, apples, and bananas. The researchers used the PLS method, MLR, and the hyper-spectral imaging technique in the visible and near-infrared to evaluate the internal attributes of fruit and vegetables.

In the national publications shown in table 10.3, most of the research involved Indian fruits, such as mangoes, bananas, apples, and pomegranates. The researchers used image processing techniques such as the *k*-means clustering technique, the blob extraction method, the template matching method, and SVM to extract the features of fruit and vegetables.

Non-destructive methods are well and good. However, these methods are expensive and skilled labor is required to perform the assessment. Moreover, usable systems are not yet available in the market. A lot of research will be required to produce friendly and usable systems.

10.8 Conclusions

In the past, the inspection, grading, and quality assessment of fruit and vegetables were done manually. The internal parameters of fruits cannot be manually tested; moreover, manual testing takes a lot of time and is inefficient. At a later stage, destructive methods were used to assess the quality of fruit and vegetables. Chemicals were used in these techniques, wastage was very high, and these techniques were unsatisfactory. Today, non-destructive methods, such as optical and NIR spectroscopy have great potential for use in the assessment of food quality. Using these techniques, various assessments such as maturity, firmness, and ripeness can be performed more effectively. Even though these techniques are costly compared to manual assessment, they are more accurate and they also save time. Hence, the future of quality assessment mainly depends on non-destructive methods, such as optical and spectroscopic techniques.

Bibliography and further reading

[1] Paliwal N, Vanjani P, Liu J-W, Saini S and Sharma A 2019 Image processing-based intelligent robotic system for assistance of agricultural crops *Int. J. Soc. Human. Comput.* **3** 191–204

[2] Jain V, Jhawar D, Saini S, Perumal T and Sharma A 2021 Human activity recognition in ambient sensing using sequential networks *Proc. Int. Conf. on Paradigms of Computing, Communication and Data Sciences* pp 493–501

[3] Lata K and Saini S 2021 A review of deep learning-based methods for cancer detection and classification *Deep Learning for Biomedical Applications* 1st ed. (Boca Raton, FL, USA: Imprint CRC Press)

[4] Cen H and He Y 2007 Theory and application of near infrared reflectance spectroscopy in determination of food quality *Trends Food Sci. Technol.* **18** 72–83

[5] Liu Y and Zhou Y 2013 Quantification of the soluble solids content of intact apples by VIS-NIR transmittance spectroscopy and the LS-SVM method *Spectroscopy* **28** 32–43

[6] Liu Y, Sun X and Ouyang A 2010 Nondestructive measurement of soluble solid content of navel orange fruit by visible-nir spectrometric technique with plsr and pca-bpnn *LWT - Food Sci. Technol.* **43** 602–7

[7] Wold S, Sjostrom M and Eriksson L 2001 Pls-regression: a basic tool of chemometrics *Chemom. Intell. Lab. Syst.* **58** 109–30

[8] Moghimi A, Aghkhani M H, Sazgarnia A and Sarmad M 2010 Vis/nir spectroscopy and chemometrics for the prediction of soluble solids content and acidity (ph) of kiwifruit *Biosyst. Eng.* **106** 295–302

[9] Roger J M and Bellon-Maurel V 2000 Using genetic algorithms to select wavelengths in near-infrared spectra: application to sugar content prediction in cherries *Appl. Spectrosc.* **54** 1313–20

[10] Sun T, Xu W, Lin J, Liu M and He X 2012 Determination of soluble solids content in navel oranges by vis/nir diffuse transmission spectra combined with cars method. *Spectrosc. Spectr. Anal.* **32** 3229–33

[11] Pissard A, Pierna J A F, Baeten V, Sinnaeve G, Lognay G, Mouteau A, Dupont P, Rondia A and Lateur M 2013 Non-destructive measurement of vitamin c, total polyphenol and sugar content in apples using near-infrared spectroscopy *J. Sci. Food Agric.* **93** 238–44

[12] Liu Y, Sun X, Zhou J, Zhang H and Yang C 2010 Linear and nonlinear multivariate regressions for determination sugar content of intact gannan navel orange by vis-nir diffuse reflectance spectroscopy *Math. Comput. Model.* **51** 1438–43

[13] Ilari J L, Martens H and Isaksson T 1988 Determination of particle-size in powders by scatter correction in diffuse near-infrared reflectance *Appl. Spectrosc.* **42** 722–8

[14] Liu Y, Sun X and Ouyang A 2010 Nondestructive measurement of soluble solid content of navel orange fruit by visible-nir spectrometric technique with PLSR and PCA-BPNN *LWT-Food Sci. Technol.* **43** 602–7

[15] Wu D, He Y and Bao Y D 2006 Fast discrimination of juicy peach varieties by Vis/NIR spectroscopy based on Bayesian-SDA and PCA *Lecture Notes in Computer Science* **vol 4113** pp 931–6

[16] Cen H, He Y and Huang M 2007 Combination and comparison of multivariate analysis for the identification of orange varieties using visible and near infrared reflectance spectroscopy *Eur. Food Res. Technol.* **225** 699–705

Shi Y, Wei W and He Z 2015 A lightweight white-box symmetric encryption algorithm against node capture for WSNs *Sensors* **15** 11928–52

[17] Cao F W D, He Y and Bao Y 2009 Variety discrimination of grapes based on visible-near reflectance infrared spectroscopy *Acta Opt. Sin.* **29** 537–40

[18] Tavakolian M S S, Silaghi F A, Fabbri A, Molari G, Giunchi A and Guarnieri A 2013 Differentiation of post harvest date fruit varieties non-destructively using FT-NIR spectroscopy *Int. J. Food Sci. Technol.* **48** 1282–8

[19] Jaiswal P, Shyam N and Bharadwaj J 2012 Non-destructive prediction of quality of intact banana using spectroscopy *Sci. Hortic.* **135** 14–22

[20] Rajkumar P, Wang N, Elmasry G, Raghavan G S V and Gariepy Y 2012 Studies on banana fruit quality and maturity stages using hyperspectral imaging *J. Food Eng.* **108** 194–200

[21] Jhuria M, Kumar A and Borse R 2013 Image processing for smart farming : detection of diseases and fruit grading *Proc. IEEE Second Int. Conf. on Image Information Processing (ICIIP-2013)* pp 521–6

[22] Pushpa B R, Tripulla K H and Meghana T K 2015 Detection and classification of fungal disease in fruits using image processing techniques *Int. J. Appl. Eng. Res.* **10** 1757–61

[23] Amodioa M L and Cegliea F March 2017 Potential of NIR spectroscopy for predicting internal quality and discriminating among strawberry fruits from different production systems *Postharvest Biol. Technol.* **125** 112–21

[24] Caramêsa E T S, Alamara P D, Poppib R J and Lima Pallone J A March 2017 Quality control of cashew apple and guava nectar by near infrared spectroscopy *J. Food Compos. Anal.* **56** 41–6

[25] Pissard A, Baeten V, Dardenne P, Dupont P and Lateur M 2018 Use of NIR spectroscopy on fresh apples to determine the phenolic compounds and dry matter content in peel and flesh *Biotechnol. Agron. Soc. Environ.* **22** 3–12

[26] Yu J, Wang H, Sun X and Huang W 2017 Parameter optimization in soluble solid content prediction of entire bunches of grape based on near infrared spectroscopic technique *J. Food Meas. Char.* **11** 1676–80

[27] Subedi P P and Walsh K B 2011 Assessment of sugar and starch in intact banana and mango fruit by SWNIR spectroscopy *Postharvest Biol. Technol.* **62** 238–45

[28] Guthrie J and Walsh K 1997 Non-invasive assessment of pineapple and mango fruit quality using near infra-red spectroscopy *Aust. J. Exp. Agric.* **37** 253–63

[29] Mendoza F, Lu R, Ariana D, Cen H and Bailey B 2011 Integrated spectral and image analysis of hyperspectral scattering data for prediction of apple fruit firmness and soluble solids content *Postharvest Biol. Technol.* **62** 149–60

[30] Zude M 2003 Non-destructive prediction of banana fruit quality using VIS/NIR spectroscopy *Fruits* **58** 135–42

[31] Ansari M Z, Deepika Minz P and Nirala A K 2012 Fruit quality evaluation using biospeckle techniques *1st Int'l Conf. on Recent Advances in Information Technology (RAIT-2012)* pp 873–6

[32] Sindhi K, Pandya J and Vegad S 2016 Quality evaluation of apple fruit: a survey *Int. J. Comput. Appl. (0975—8887)* **136** 32–6

[33] Jadhav R S and Patil S S 2013 A fruit quality management system based on image processing *IOSR J. Electron. Commun. Eng. (IOSR-JECE)* **8** 1–5

[34] Bhargava A and Bansal A 2018 Fruits and vegetables quality evaluation using computer vision: a review *J. King Saud Univ.—Comput. Inform. Sci.* **33** 243–57

[35] Bhangea M and Hingoliwalab H A 2015 Smart farming: pomegranate disease detection using image processing *Second Int. Symp. Comp. Vision Internet (VisionNet'15), Procedia Comp. Sci.* **58** 280–8

[36] Dubey S R and Anand Siggh Jalal A S 2012a Detection and classification of apple fruit diseases using complete local binary patterns *3rd IEEE Int. Conf. Comp. Commun. Technol. (ICCCT-2012)* p 3

[37] Pandey C, Kumar Sethy P, Biswas P, Behera S K and Khan M R 2020 Quality evaluation of pomegranate fruit using image processing techniques *Int. Conf. on Communication and Signal Processing (ICCSP)* pp 38–40

[38] Payel G and Sunil C K 2014 Quality analysis of pomegranate by non-destructive soft x-ray method *J. Food J Proces. Technol.* **5** 1–6

[39] ElMasry G, Wang N, ElSayed A and Ngadi M 2007 Hyperspectral imaging for nondestructive determination of some quality attributes for strawberry *J. Food Eng.* **81** 98–107

[40] Huang H, Yu H, Xu H and Ying Y 2008 Near infrared spectroscopy for on/in-line monitoring of quality in foods and beverages: a review *J. Food Eng.* **87** 303–13

[41] McClure W F 1994 Near-infrared spectroscopy—the giant is running strong *Anal. Chem.* **66** A43–53

[42] Nicolaï B M, Beullens K, Bobelyn E, Peirs A, Saeys W, Theron K I and Lammertyn J 2007 Nondestructive measurement of fruit and vegetable quality by means of NIR spectroscopy: A review *Postharvest Biology and Technology* **46** 99–118

[43] Roger J M and Bellon-Maurel V 2000 Using genetic algorithms to select wavelengths in near-infrared spectra: application to sugar content prediction in cherries *Appl. Spectrosc.* **54** 1313–20
[44] Peiris K H S, Dull G G, Leffler R G and Kays S J 1998 Near-infrared spectrometric method for nondestructive determination of soluble solids content of peaches *J. Am. Soc. Hortic. Sci.* **123** 898–905
[45] DLP® NIRscan™ Nano EVM User's Guide by Texas Instruments
[46] DLP Technology for Spectroscopy. White Paper DLPA048A–February 2014–Revised August 2016.

IOP Publishing

Advances in Image and Data Processing using VLSI Design, Volume 1
Smart vision systems
Sandeep Saini, Kusum Lata, Abhishek Sharma and G R Sinha

Chapter 11

Design and implementation of processors for secure image processing applications

M Vardhana, Roopashree and Pavanalaxmi

11.1 Introduction

Microprocessors and microcontrollers have the advantages of single-cycle multiply-accumulate instructions and special addressing modes, which have enabled them to dominate the field of digital signal processing. Digital signal processors appeared on the market in the early 1980s. They have been the key empowering innovation for an enormous number of electronics products in, for example, the communications, automotive, instrumentation, and military fields.

As integrated circuit (IC) technologies have become a pressing need, the development of controllers and processors has received considerable attention in the last few years, and much research work has attempted to improve their speed and efficiency. This has interested us in the development of a programmable microcontroller with signal processing operations that can reduce code complexity.

Existing microcontrollers and processors support simple arithmetic and logical operations along with a basic instruction-set architecture. The ALUs of existing systems use shift and add methods for their multiplication operations; the TMS320C54XX uses an array multiplier for the multiplication operation and barrel shifters for shift operations. The clock cycles of existing processors and controllers vary from one to four; the TMS320C54XX has dedicated hardware for the multiplier and the accumulator and also supports bit-reversed addressing modes.

Existing systems do not have dedicated hardware for multiplication, division, etc., nor do they have built-in encryption/decryption units. Conventional processors implement signal processing operations such as FFT and convolution as program modules and do not use dedicated hardware to increase the program memory. After a survey of many related papers, we concluded that the ancient Vedic mathematical techniques have greater computational speed than other methods.

doi:10.1088/978-0-7503-3919-3ch11

The following sections of the chapter list some of the research based on Vedic mathematics. Brief details of Vedic mathematics are given in section 11.3. The remaining sections of the chapter describe the design of the proposed system along with the simulation results and the conclusions.

11.2 Literature survey

Chidgupkar *et al* [1] described the use of Vedic algorithms in digital signal processing and also stressed the importance of Vedic sutras for high-speed computation. The authors attempted to save an appreciable amount of the processing time required to perform multiplication routines using Vedic algorithms. A Vedic multiplication algorithm was implemented on 8085/8086 microprocessors. The authors reported that the use of Vedic algorithms was beneficial in DSP applications. Many researchers are exploring Vedic algorithms in detail to understand their applicability in various engineering fields.

Itawadiya *et al* [2] described a new method for computing high-speed convolution based on Vedic mathematics. The fast computation of digital signal processing operations was carried out using the Vedic Urdhva triyagbhayam multiplication sutra. The Vedic sutra was carried out using a graphical user interface (GUI), and all digital signal processing operations were performed within the same GUI window. The digital signal processing operations based on this sutra reduced the processing time, compared to the time required by conventional functions.

Kareem *et al* [3] proposed a new technique for multiplication using a modified Vedic sutra. A 16 × 16 multiplier integrated circuit was designed using an available electronics design automation (EDA) tool. A comparison of various multiplier architectures was presented; the comparison revealed that the proposed multiplier was proven to be better than conventional multipliers, using power, area, and delay as the factors of interest.

Kumar [4] proposed a design for a FFT circuit based on the Vedic sutra. The authors used the Urdhva tiryakbhyam and Nikhilam sutras to obtain high throughput. Using Vedic sutras, the number of adders and multipliers required to perform a FFT was reduced. The reduction in gate numbers and circuit size reduced the area required, reducing delay and, in turn, increasing the speed of the FFT.

Lomte [5] presented a direct method for the convolution and deconvolution of information, which is the basic operation involved in various digital signal processing applications. The Urdhva triyagbhyam algorithm is used to find the discrete convolution output for two finite-length sequences. The speed of the deconvolution can be improved using the same algorithm. The presented design speeds up computation with no change in the area.

Kumar and Charishma [6] implemented a Vedic multiplier design on a Spartan kit. The simulation results were computed using 8-bit, 16-bit, and 32-bit multipliers. The results obtained were compared with the Karatsuba algorithm, which showed that the design of the Vedic multiplier improved efficiency and speed.

Nivas and Kayalvizhi [7] presented a new architecture for a Vedic multiplier with reduced power consumption. Three different cases were considered; a simulation was performed using the Verilog hardware description language (HDL) and synthesis was performed using the Synopsys design compiler. This power-efficient Vedic multiplier can easily be utilized in any digital signal processing application or in any digital signal processor in which power consumption is a neglected factor.

Pradhan [8] proposed a high-speed multiplier circuit that was implemented on a Spartan device. The proposed multiplier was simulated using the Xilinx ISE software by writing very high-speed integrated circuit HDL (VHDL) code and a test bench was written by considering five sets of data. The synthesis results were compared with other multiplication methods, and showed an improvement in speed.

Jain *et al* [9] used the Nikhilam sutra and the Parvartya sutra in the design of an optimized binary division architecture. This work examined these two division methods and their application to deconvolution. The coding of the proposed algorithm coding was carried out in Verilog, and a simulation was carried out using the Xilinx ISE design suite version 14.2.

11.3 Vedic mathematics

Vedic Mathematics is the title given to the archaic system of Indian mathematics that originated from the Vedas somewhere between the years 1911 and 1918 by Sri Bharati KrsnaTirthaji (1884–1960). As indicated by his exploration, all of this mathematics depends on sixteen sutras. These formulae describe how the mind normally works and are accordingly of extraordinary assistance in guiding students to the appropriate techniques for calculation.

Maybe the most salient element of the Vedic background is its coherence. Rather than a hotchpotch of inconsequential procedures, the entire subject is wonderfully interconnected and coherent: the whole multiplication procedure, for example, can readily be reversed to permit one-line divisions and the straightforward squaring strategy can be switched to give one-line square roots. In addition, these can readily be comprehended as a whole. This coherent quality is extremely satisfying; it makes arithmetic simple and pleasant and permits progress. In the Vedic framework, 'troublesome' matters can regularly be tackled faster by Vedic techniques. These salient and wonderful schemes form a complete arrangement of science that is more deliberate than an advanced 'framework.' Vedic Mathematics demonstrates the soundness and cohesion of the design of mathematics, and its approaches are common, direct, and simple.

The truthfulness of Vedic Mathematics implies that estimations can be accomplished intellectually. There are numerous benefits in utilizing an adaptable, conceptual background. Learners can improve their techniques; they are not limited to the unique 'right' strategy. This encourages more innovative, fascinated, and keen learners. The examination is being completed in numerous spaces remembering the impacts of learning Vedic Mathematics for youngsters; creating new, incredible, yet simple uses of the Vedic sutras in calculation, analytics, registering, and so forth; in any case, the genuine magnificence and sufficiency of Vedic Mathematics cannot be completely understood without experiencing it.

11.3.1 Vedic sutras

'Vedic' originates from 'Veda' which implies a repository of all information. Vedic mathematics is essentially founded on 16 sutras (or adages) that manage different parts of science, such as number juggling, variable-based math, calculation, and so forth.

These techniques and thoughts can be straightforwardly applied to geometry, plain and circular calculations, conics, analytics, and applied arithmetic of different sorts. As mentioned previously, every one of these sutras was fairly recently recreated from antiquated Vedic writings. Many sub-sutras were additionally found at the same time, which are not discussed here. The magnificence of Vedic arithmetic lies in the way that it reduces the generally lumbering indirect computations in regular mathematics to an extremely basic form. This is possible because the Vedic formulae are professed to be founded on the same principles as those on which human thoughts work.

11.3.2 Urdhva tiryagbhyam

These sutras are the essential ones, which are relevant to all instances of duplication [3]. They are exceptionally tiny and concise, comprising just one consolidated word that signifies 'vertically and transversely'; for example, the principal part of the multiplicand and the main part of the multiplier are duplicated using the 'vertical and sideways' technique [4, 5]. The vertical and transverse duplication system is also called the cluster increase method. Figure 11.1 shows the model used for multiplicative utilization of the Urdhva sutra.

11.3.3 Paravartya

The Paravartya vedic sutra is used for division; the steps involved in the process of division are as follows: given a dividend and a divisor, the first step is to compare the divisor with the base number to find out which is greater. A positive difference arises if the divisor is greater; otherwise, a negative difference results. The new divisor for

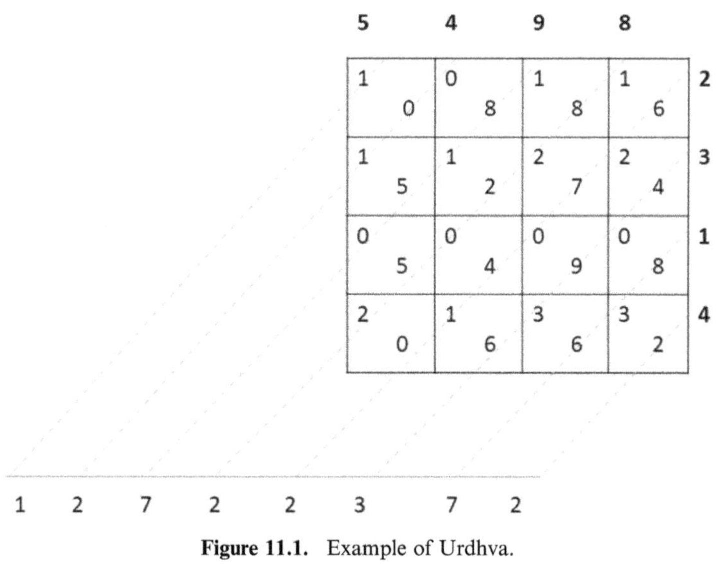

Figure 11.1. Example of Urdhva.

further steps is the difference obtained in the previous step. We bring down the first digit of the dividend, and multiply the first digit by the difference. The result is written in the next line of the dividend shifted one bit to the right. We perform addition to get the second bit and write it down. We repeat the operation until we reach the last digit of the dividend. The quotient and the remainder can be obtained by following these steps. Figure 11.2 shows the application of this sutra for BCD division [10–12].

11.4 Proposed system design and architecture

This system presents the design and implementation of an eight-bit Vedic microcontroller, which performs basic arithmetic operations using ancient Vedic mathematical techniques. A microcontroller is a programmable device with essential peripherals. Microcontrollers and microprocessors are complex circuits that perform precise work based on their instructions or programs. A fixed set of programs is placed in the read-only memory of a microcontroller, and those programs regulate the functioning of the machine. The programs installed do not change over time. A set of instructions written in a precise order for a system to solve a particular task is known as a program, and a set of programs is called software. A program saved in system memory in binary form is composed of machine-level instructions. The evolution of these programmable devices has been tremendous. The impact or application of programmable devices is not well known to their users, even when these devices have evolved to a significant degree. Microcontrollers are used in most electronic devices, from home appliances to industrial applications and even space applications, etc. The performance of the system

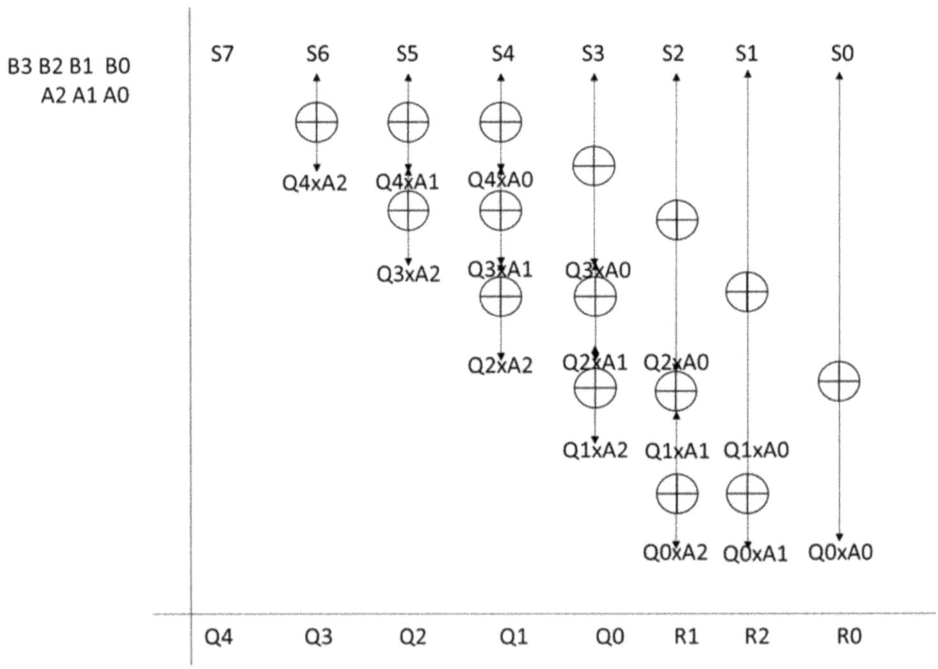

Figure 11.2. Example of BCD division using Paravartya.

depends upon the efficiency of the controller used. Optimization of the controller leads to the optimization of the entire system. The speed and efficiency of the system depend upon the speed and efficiency of the microcontroller used. Thus, the design of high-speed microcontrollers is a major area of interest. The basic building blocks of microcontrollers are the ALU, random-access memory (RAM), read-only memory (ROM), the stack, the registers, and the input–output (IO) ports. The ALU is the core of a microcontroller; it performs arithmetic and logical calculations. This chapter proposes the design and use of an eight-bit microcontroller whose ALU performs computations based on the Vedic sutras. The proposed controller consists of a 256 × 8 RAM, a control unit, an 8 × 8 stack, working registers R0–R7, a flag indicator, and accumulators A, B, C, and D. All 256 bytes of RAM are byte addressable. RAM locations 00 to 03 are allocated to registers A, B, C, and D, whereas locations 04 to 07 are allocated to registers R0–R3. The stack pointer is used to point to the default stack location 0Ch. The stack can be assessed by the instructions PUSH and POP. The remaining RAM locations are general-purpose. These RAM locations are byte addressable and only indirect addressing modes are allowed. The ALU of the proposed Vedic microcontroller is built using Vedic mathematical techniques, which are much faster than conventional techniques. A microcontroller is one of the basic components of most control units and hence the efficiency of the system mainly depends on the controller's actions. Thus, the design of a high-speed microcontroller is the main area of interest. In this system, an efficient high-speed microcontroller is designed and implemented using very large-scale integration (VLSI) using an EDA tool.

11.4.1 Design overview

The system design has dedicated address and data buses and an instruction queue which holds the instruction bytes. The current instruction in the instruction queue is decoded by the instruction decoder which generates the opcode. The ALU consists of an eight-bit adder, shift registers, a multiplier, and a divider. To improve the efficiency of the system, we developed dedicated hardware which performs various signal processing operations. Verilog code was written for all the modules and combined to develop the proposed controller. The controller is programmed using the proposed compiler. The compiler converts assembly-level language to machine code. Thus, the generated hex file is stored in the ROM of the controller and is used for processing. The register window displays the current values of the accumulators and general-purpose registers, along with the flag registers (figure 11.3).

11.4.2 Fundamental building blocks

11.4.2.1 Multiplier unit

Urdhva tiryagbhyam (which means vertical and crosswise) is the Vedic sutra used for the multiplication of large numbers. It can be used to multiply any two large numbers without much calculation. We applied this sutra in the development of an 8 × 8-bit multiplier unit that is part of the ALU of the proposed controller, the VMPV_008K. The instruction MUL is used for multiplication, in which the final product, which is 16 bits long, is saved in accumulators A and B.

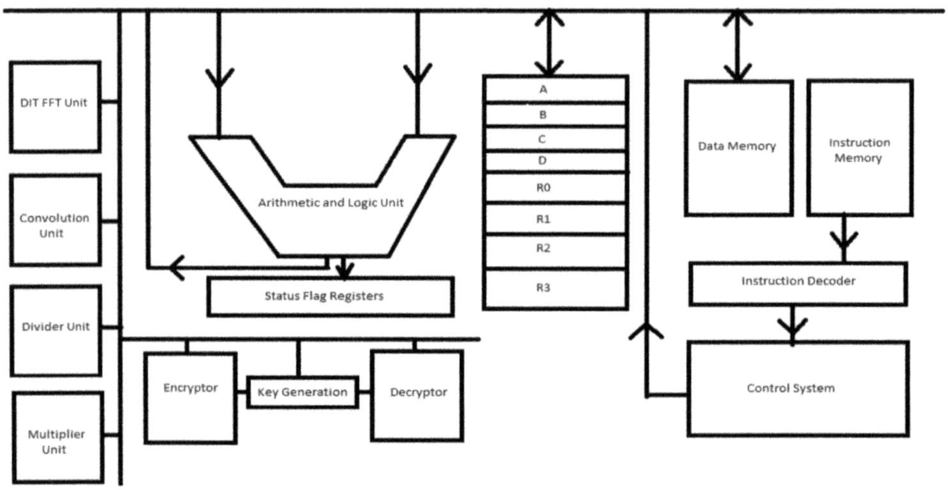

Figure 11.3. Architecture of the proposed system.

11.4.2.2 Divider unit

For the proposed controller, we employed a Vedic sutra called Paravartya in the development of the divider unit of the ALU. The eight-bit dividend is saved in A, the four-bit divisor is saved in B, and when the instruction DIV is executed, the division of the values in registers A and B takes place; the quotient is stored in A and the remainder in B.

11.4.2.3 Convolution unit

Convolution plays a very significant part in current computational systems. The convolution process relates the input to the output of a system whenever the impulse response of a system is known. For a continuous-time system, the output can be evaluated using a convolution integral, and for a discrete system, by the convolution sum. The convolution sum is a scientific procedure that relates the output [yn], of a linear time invariant (LTI) system, to its input $x[n]$ and its impulse response $h[n]$. This procedure is represented by CONV ($x[n]$, $h[n]$).

In the proposed controller, convolution is implemented using Vedic techniques in which the multiplication operation present in the convolution is replaced by the proposed Vedic multiplier.

Convolution is the sum of the product of shifted samples. Thus, by increasing the speed of the multiplication and by using pipelining, the speed of the convolution unit can be greatly increased.

In the proposed controller, the instruction CONV performs the convolution operation. The input samples, which are both 8 bits, are stored in the memory locations from 32H to 35H and the 8-bit coefficients of the impulse response are stored in the memory locations from 36H to 39H.

On executing the instruction CONV, the convolution of the input samples and the coefficients of the impulse response are calculated and stored in the memory locations from 3AH to 49H. Two memory blocks are allocated for each sample of the output.

11.4.2.4 Fast Fourier transform unit

Fast Fourier transform (FFT) is utilized to translate signals from the time domain to the frequency domain. Since the analysis of some operations becomes difficult in the time domain, the frequency domain is often selected for signal processing; FFT is a tool that transforms a signal between one domain and the other. FFT algorithms can be classified into decimation in time (DIT) FFT and decimation in frequency FFT (DIFFFT).

In the proposed controller, we employ the DIT FFT algorithm to calculate the DFT of the signal. The input samples are represented in the Q0 notation and the twiddle factors are represented in the Q7 notation. The final output is in the Q7 notation.

The memory addresses from 4AH to 59H are allocated for the FFT input samples; the real and imaginary parts are stored in separate memory locations. The transformed output data are stored in a memory location from 5AH to 69H.

11.4.2.5 Encryption/decryption unit

Security is essential when communicating through untrusted media. It is essential to keep messages that are being communicated secret and confidential. Encryption is the solution to this problem. A cryptographic algorithm can be classified as either symmetric or asymmetric, according to the number of keys used for encryption/decryption. This controller employs the advanced encryption standard (AES), which is a symmetric cryptographic algorithm. It uses simplified AES (SAES), in which the key and plaintext sizes are 16 bits.

The instruction ENCRYPT encrypts the plaintext saved in A and B, the key is saved in registers R0 and R1, and the 16-bit ciphertext is saved in C and D. The instruction DECRYPT decrypts the ciphertext saved in C and D, the key is stored in registers R0 and R1, and the 16-bit plaintext is stored in accumulators C and D.

11.4.3 Instruction set architecture

The instruction set can be broadly classified into the following groups:

- Data-transfer instructions: MOV, PUSH, POP
- Loop instructions: BEGIN-LOOP
- Arithmetic instructions: ADD, SUBB, MULL, DIV
- Logical instructions: XOR, AND, OR, SRL
- Signal processing instructions: CONV, FFT, ENCRYPT, DECRYPT

11.4.4 Software specification and design flow

The proposed VMPV_008K processor is implemented using an EDA tool, and the compiler design is completed. In the first stage, Verilog code is written for each module and tested for the same. The tested modules are integrated to build the final processor. The Verilog code is simulated using a Verilog simulator and the output is observed for a sample test program. A compiler is designed, which converts the high-level language to machine code. A backend tool is used to design the IC of the proposed processor. The floor planning is the first step in the IC design; the power planning is then completed, and horizontal and vertical power strips are selected.

The basic cells are placed on the design and routing is performed. In the final step of the IC design, a Graphic Design System (GDS) II file is generated.

11.5 Simulation results

The test bench is written using Verilog code and simulated using code simulators; the RTL schematic is obtained. The following figures show the RTL schematic of the building blocks of the processor (figures 11.4–11.14).

11.5.1 Integrated circuit design

The IC design for the proposed system is produced using a backend tool. The netlist is generated and imported into a layout editor. The following figures show the IC design of the proposed processor (figures 11.15–11.21).

11.5.2 Compiler design

We produced a compiler for the proposed system. Regular expressions were created according to the instruction templates and expression matching was performed. Once the compilation was complete, the bit file to be added to the system memory was generated by the proposed processor. The following figures show the design and an overview of the compiler (figures 11.22 and 11.23).

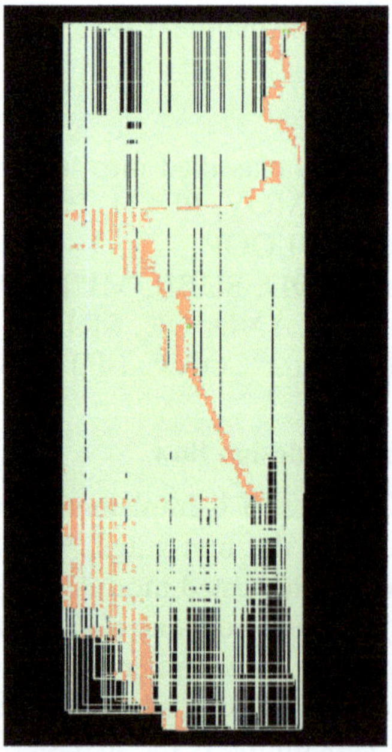

Figure 11.4. Main processor.

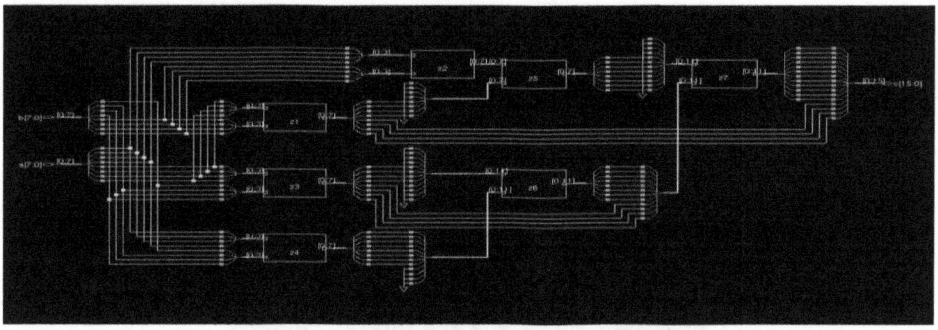

Figure 11.5. Vedic multiplier.

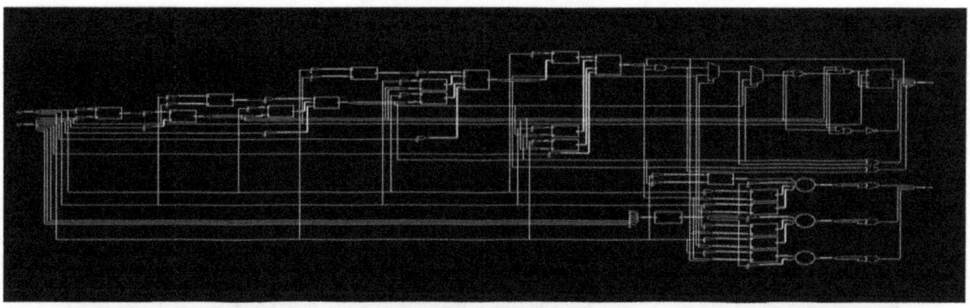

Figure 11.6. Vedic divider.

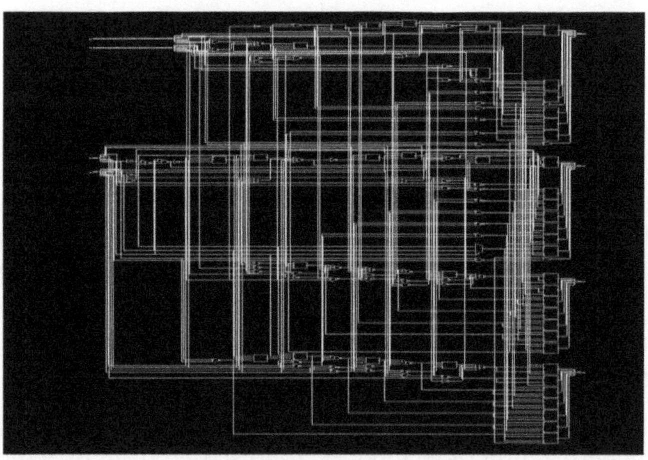

Figure 11.7. 8 point FFT.

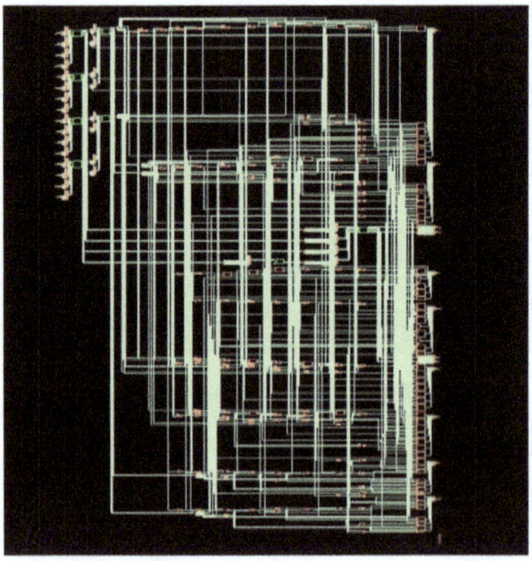

Figure 11.8. 8 point FFT.

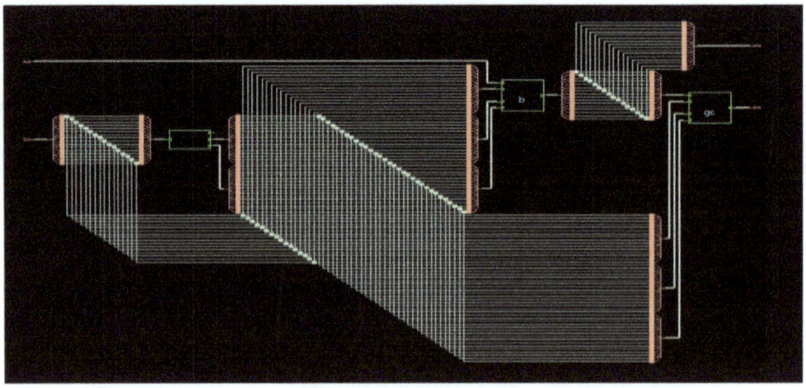

Figure 11.9. AES unit.

11.5.3 Performance analysis

A synthesis of the proposed system was performed and synthesis reports were generated. The power consumed, the timing delay, and the area consumed by each module of the proposed processor are presented in the following section. Tables 11.1, 11.2, and 11.3 show the power, area, and timing details of the proposed processor.

11.6 Applications

The techniques used in the diagnosis of disease have undergone great changes, resulting in a modernization of techniques, followed, in turn, by a modernization of devices. A revolutionary change has occurred in the healthcare industry because of

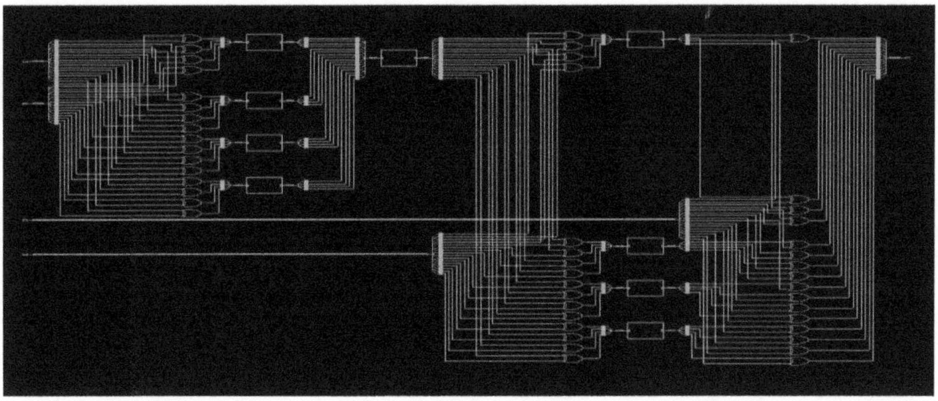

Figure 11.10. Encryption unit.

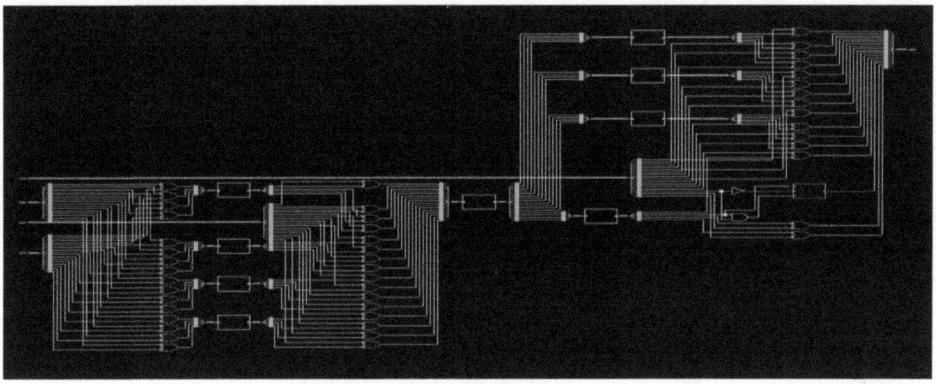

Figure 11.11. Decryption unit.

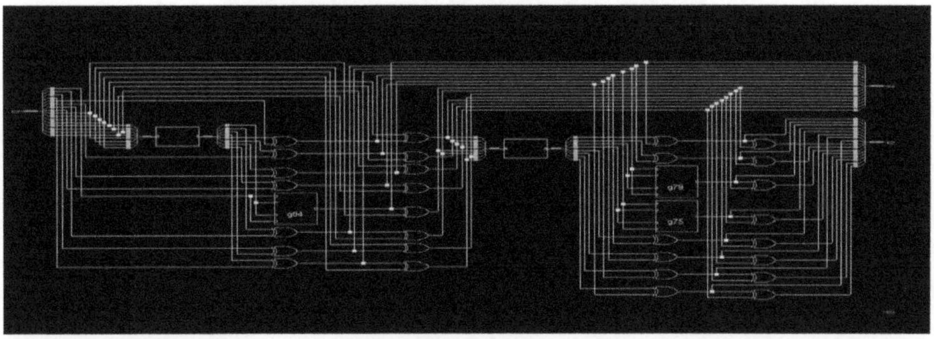

Figure 11.12. Key generation unit.

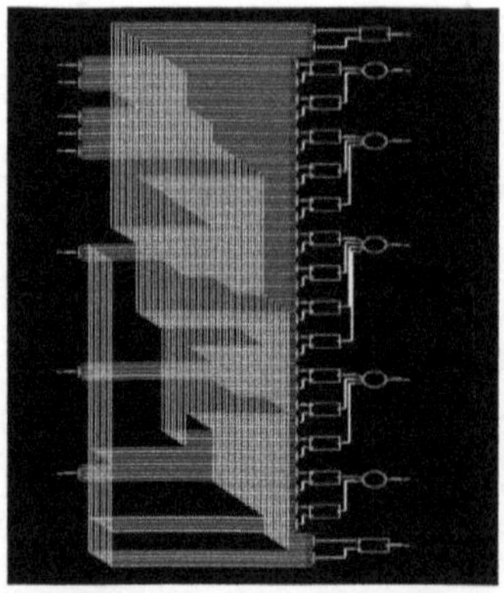

Figure 11.13. Vedic convolution block.

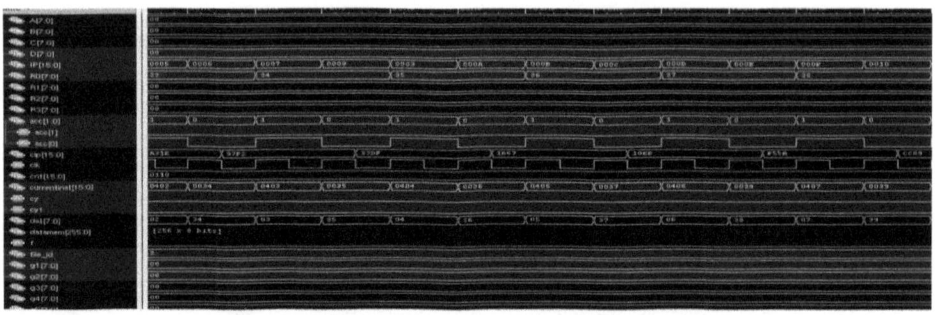

Figure 11.14. Simulation result for a sample program.

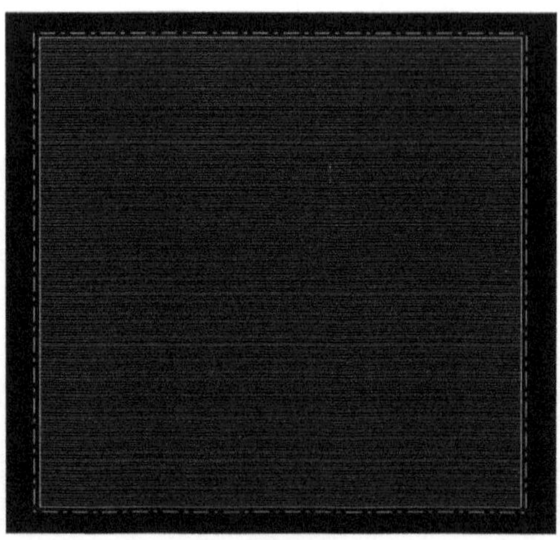

Figure 11.15. Floor planning.

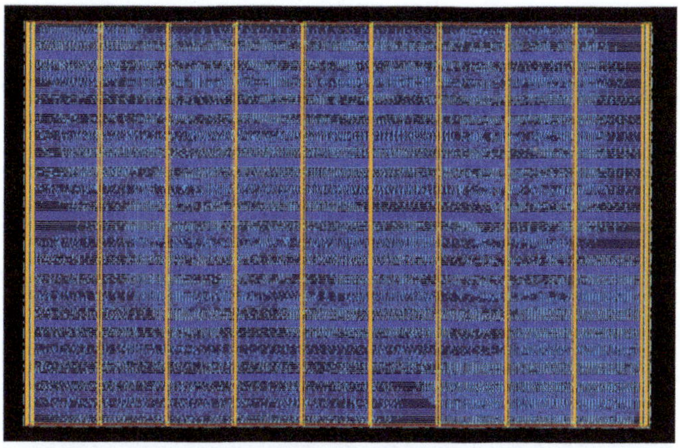

Figure 11.16. Power planning and placement.

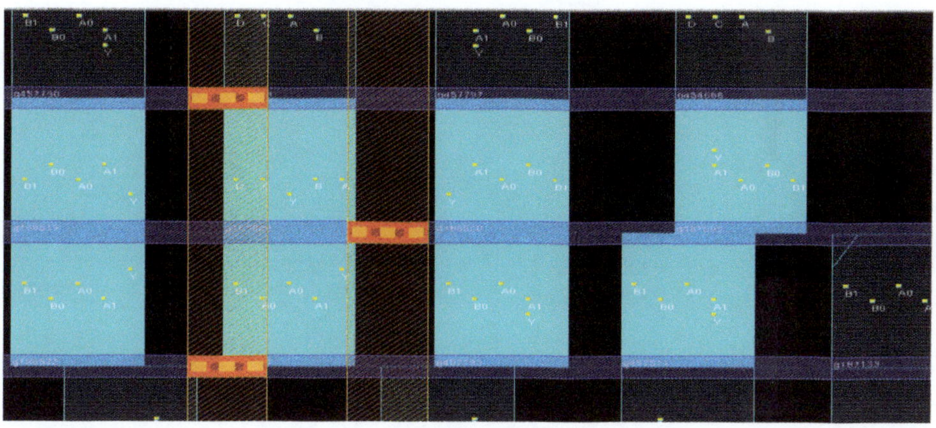

Figure 11.17. Placement of the primitive cells.

the use of medical image processing. Medical image processing is an advanced application of digital image processing. Medical images provide information embedded in maximum-resolution slices of captured data, which increase the size of the image. These images can be used in interactive telemedicine services, wherein image exchange takes place between the patient and physician. In such applications, the network resources used have limited bandwidth for image transmission. Therefore, this type of application requires a highly robust compression algorithm with a good response rate. The response rate is a measure of how fast an algorithm can encode a stream of medical data and generate a reconstructed image without losing significant information of clinical value. Most algorithms use multipliers. The Vedic Mathematics-based multiplier performs compression faster than conventional schemes. Vedic Mathematics-based DSP operations reduce the processing time required. Convolution is a basic technique used for edge detection in many digital

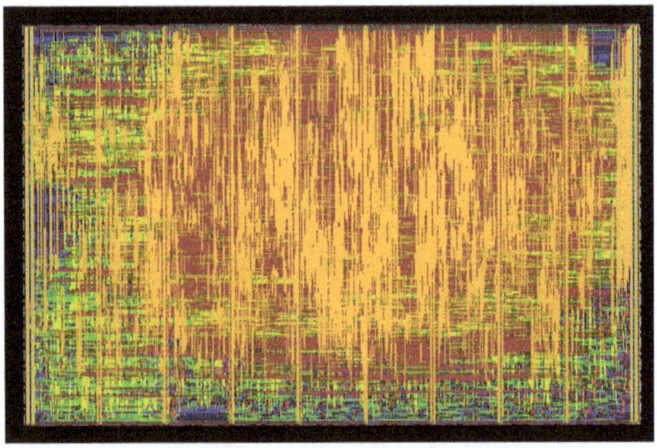

Figure 11.18. The finished VLSI chip of the proposed processor.

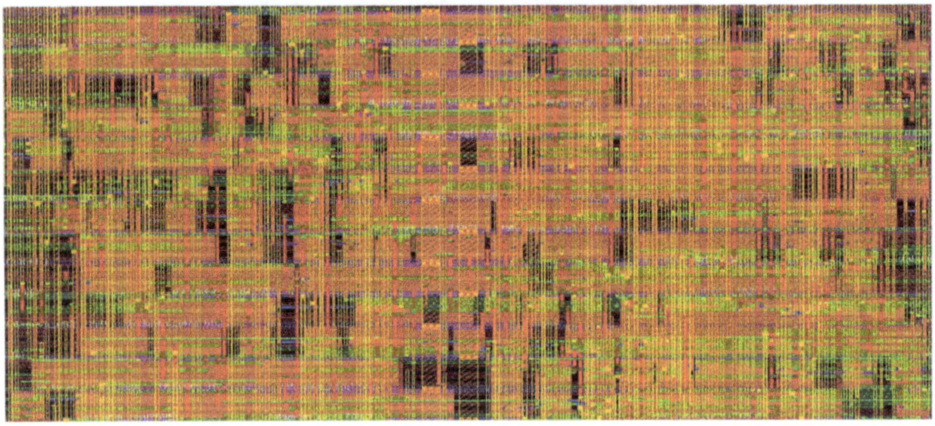

Figure 11.19. Magnified view of the processor.

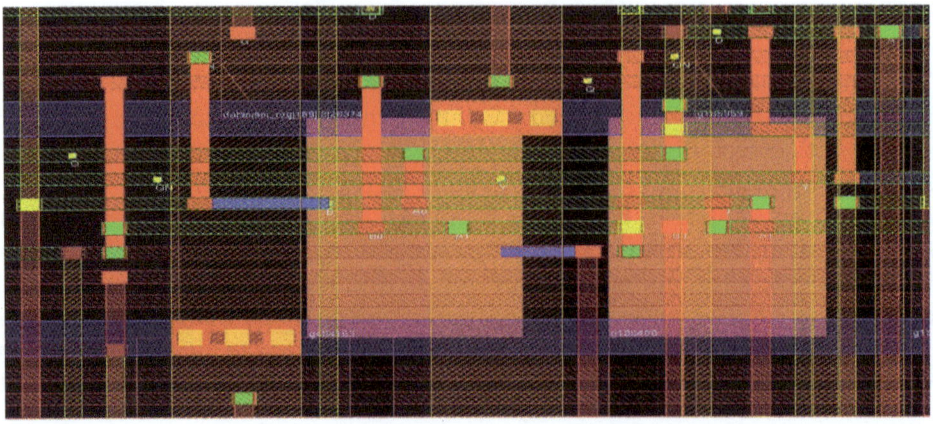

Figure 11.20. Magnified view showing the data memory.

```
Stream Out Information Processed for GDS version 3:
Units: 2000 DBU

Object                                   Count
------------------------------------------------
Instances                                  717

Ports/Pins                                  64
    metal layer Metal2                      29
    metal layer Metal3                      25
    metal layer Metal4                       5
    metal layer Metal5                       4
    metal layer Metal6                       1

Nets                                      4263
    metal layer Metal1                     437
    metal layer Metal2                    2232
    metal layer Metal3                    1402
    metal layer Metal4                     175
    metal layer Metal5                      16
    metal layer Metal6                       1

    Via Instances                         5428

Special Nets                                99
    metal layer Metal1                      85
    metal layer Metal2                       6
    metal layer Metal5                       4
    metal layer Metal6                       4

    Via Instances                          399

Metal Fills                                  0

    Via Instances                            0

Metal FillOPCs                               0

    Via Instances                            0
```

Figure 11.21. Generated GDS II file.

processing applications. In this chapter, we considered a fast method for DSP operations based on ancient Vedic mathematics.

11.7 Results and conclusions

In this chapter, we proposed a new microcontroller based on ancient Vedic mathematical techniques. The system was modeled and simulated using a Verilog simulator. The IC design was completed using a backend tool. The clock frequency was 100 MHz, which is much greater than those of conventional controllers and can be increased further. The application of Vedic mathematics for computation greatly increases the efficiency of the system.

The proposed system supports all the basic addressing modes and instruction sets and also supports special instructions for signal processing. Vedic sutras are used for high-speed computation and the system was developed using Vedic mathematical techniques.

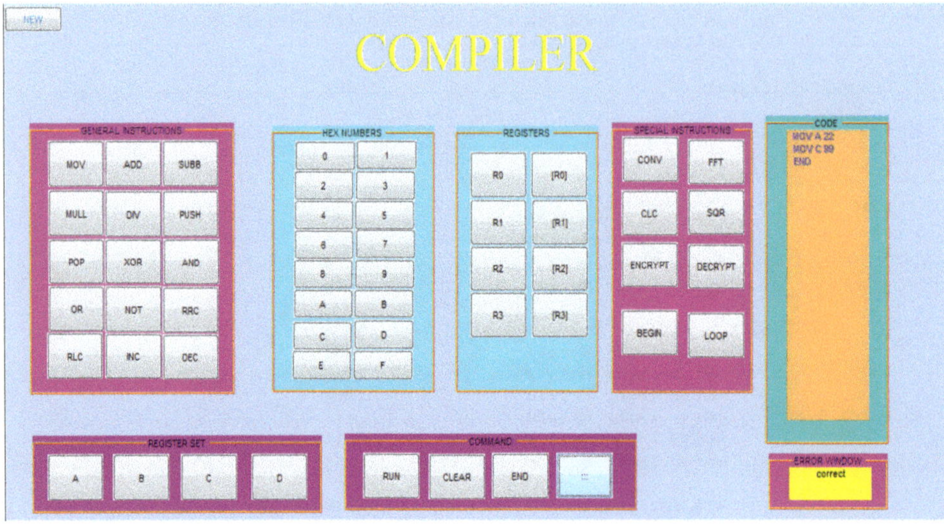

Figure 11.22. Sample code.

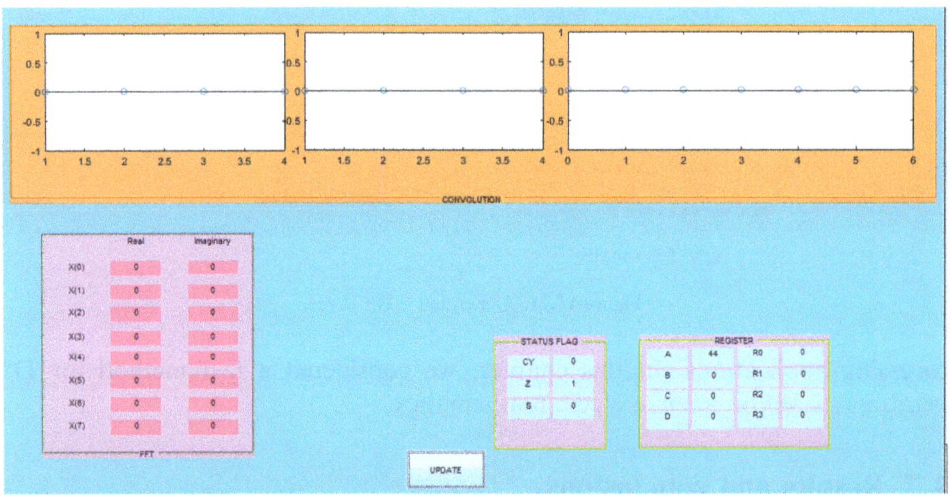

Figure 11.23. Register window.

The proposed system has built-in dedicated hardware for processing, which greatly reduces the time and memory requirements.

Using Verilog, code for the proposed system was written and synthesized and a performance analysis was completed. The physical IC design of the proposed system was completed and a GDS II file was generated.

A compiler was designed and implemented for the proposed system, which was used to convert the assembly-level language into a hex file that could be installed on the proposed system.

Table 11.1. Power details.

Module name	Leakage (nW)	Dynamic (nW)	Total (nW)
Convolution	4823.345	9281849.873	9286673.218
Decryptor	137.458	238629.737	238767.195
Encryptor	105.207	132733.403	132838.609
FFT	737.735	2706870.113	2707607.848
Divider	64.053	113753.185	113817.238
Multiplier	68.323	131668.319	131736.641

Table 11.2. Area details.

Module name	Instances	Area
Convolution	3335	104831.496
Decryptor	104	2757.586
Encryptor	80	2118.917
FFT	1549	24136.358
Divider	77	1696.464
Multiplier	37	1417.046

Table 11.3. Timing details.

Module name	Delay (ps)
Convolution	8104
Decryptor	2039
Encryptor	1464
FFT	4491
Divider	5652
Multiplier	3846

The proposed system makes significant use of Vedic mathematical techniques for high-speed calculations; it demonstrated the effective use of Vedic sutras and a programmable Vedic controller for signal processing.

The system has: (i) built-in signal processing operations; (ii) basic addressing modes; (iii) dedicated hardware to compute multiplication and division based on ancient Vedic mathematical techniques; (iv) built-in dedicated hardware to perform signal processing operations, such as FFTs, convolution, etc.; (v) support for symmetric-key cryptography.

The proposed design supports all the basic arithmetic and logical operations along with some special instructions for signal processing. The proposed processor has dedicated hardware for the computation of some complex operations, such as multiplication and division. Unlike other processors, this processor does not require pages of code to perform complex signal processing operations, such as convolution and FFTs. Dedicated hardware units are used to perform complex signal processing operations, greatly reducing the time required for execution and increasing the speed of the processor. The design was verified and fault analysis was carried out for a sample module of the processor. The design was synthesized, and the design netlist was imported into a backend tool for the physical design of the proposed processor chip. A GDS II file was generated, which was given to the foundry for use in the IC manufacturing process. This system demonstrates the performance of the proposed processor with respect to speed, area, and power. The clock frequency is 100 MHz, which is much higher than those of conventional processors and can be increased further. The application of Vedic mathematics to computation greatly improves the efficiency of the system.

Bibliography

[1] Chidgupkar P D and Karad M T 2004 The implementation of vedic algorithms in digital signal processing *Global J. Eng. Educ.* **8** 153–8

[2] Itawadiya A K, Mahle R, Patel V and Kumar D 2013 Design a DSP operations using vedic mathematics *2013 Int. Conf. on Communication and Signal Processing (Piscataway, NJ)* (IEEE) pp 897–902

[3] Kareem A and Kumar P 2016 VLSI implementation of high speed-low power-area efficient multiplier using modified vedic mathematical techniques *Recent Pat. Comput. Sci.* **9** 216–21

[4] Kumar A, Raman A, Sarin R K and Khosla A 2010 Small area reconfigurable FFT design by Vedic Mathematics *2010 2nd Int. Conf. on Computer and Automation Engineering (ICCAE)* 5 *(Piscataway, NJ)* (IEEE) pp 836–8

[5] Lomte R K and Bhaskar P C 2011 High speed convolution and deconvolution using urdhvatriyagbhyam *2011 IEEE Computer Society Annual Symp. on VLSI (Piscataway, NJ)* (IEEE) pp 323–4

[6] Kumar G and Charishma V 2012 Design of high speed vedic multiplier using vedic mathematics techniques *Int. J. Sci. Res. Pub.* **2** 1–5

[7] Kayalvizhi N and Sree Nivas A 2012 Implementation of power efficient vedic multiplier *Int. J. Comp. Appl.* **43** 21–4

[8] Pradhan M and Panda R 2014 High speed multiplier using Nikhilam Sutra algorithm of Vedic mathematics *Int. J. Electron.* **101** 300–7

[9] Jain S, Pancholi M, Garg H and Saini S 2014 Binary division algorithm and high speed deconvolution algorithm (based on ancient Indian vedic mathematics) *2014 11th Int. Conf. on Electrical Engineering/Electronics, Computer, Telecommunications and Information Technology (ECTI-CON) (Piscataway, NJ)* (IEEE) pp 1–5

[10] Jain S and Saini S 2014 High speed convolution and deconvolution algorithm (Based on Ancient Indian Vedic Mathematics) *2014 11th Int. Conf. on Electrical Engineering/ Electronics, Computer, Telecommunications and Information Technology (ECTI-CON)* pp 1–5

[11] Saini S, Mahajan A and Mandalika S B 2010 Implementation of low power FFT structure using a method based on conditionally coded blocks *2010 IEEE Asia Pacific Conf. on Circuits and Systems* pp 935–8

[12] Gupta Y, Garg L, Khandelwal S, Gupta S and Saini S 2013 Design of low power and high speed multiplexer based thermometer to gray encoder *2013 Int. Symp. on Intelligent Signal Processing and Communication Systems* pp 501–4

Chapter 12

Low-power modified phase-locked loop using AVLS technique for biomedical applications

B S Premananda, Maria Bency, Shwetha R Prasanna and G P Shreya

The power consumption of phase-locked loops (PLLs) is high because of their high switching rates. For biomedical applications, PLLs should consume minimal amounts of power. The objectives of this chapter are to modify the existing PLL architecture and incorporate power reduction techniques to minimize PLL power consumption. The subcomponents of a PLL, namely the phase detector, charge pump, loop filter, voltage-controlled oscillator, and frequency divider are individually designed and optimized for power and then integrated to form a PLL. Two different phase-detector circuits are considered: a linear phase detector and a phase frequency detector. In the loop filter design, metal–oxide–semiconductor field-effect transistor (MOSFET) are used instead of resistors and capacitors. An adaptive voltage level to supply (AVLS) is incorporated into the proposed PLL circuits to minimize the power consumption. The proposed PLL circuits are implemented using 180 nm complementary metal–oxide–semiconductor (CMOS) technology in Cadence Virtuoso and simulated in Cadence Spectre with a supply voltage of 1.8 V. The performances of reference circuits and the proposed PLL circuit are analyzed with respect to power consumption. The results confirm that the proposed PLL circuit offers a significant reduction in power consumption compared to existing PLL architectures and operates at a higher frequency (4 GHz).

12.1 Introduction

Phase-locked loops (PLLs) are employed in a wide range of applications. Due to this circuit's adaptability and the rise of the digital domain, it is now employed in communications subsystems for carrier recovery, clock recovery, modulation, demodulation, frequency synthesis, and frequency generation as well as in biomedical devices. With the use of CMOS technologies and the increase in the

doi:10.1088/978-0-7503-3919-3ch12

integrated density of microelectronic components, microelectronic technology has become more realistic and demanding.

As biomedical implants are portable, they must use very little power. Because batteries have a finite amount of energy, such implants must be designed for ultra-low power consumption. In biomedical implantable devices, the PLL is the most important functional unit; it provides a steady clock signal, but it is responsible for a significant amount of power utilization. The PLL is an integral part of electronic implanted biomedical devices. The first stage in creating a low-power biomedical implant is to create a low-power PLL [1].

The block diagram of the PLL depicted in figure 12.1 consists of a phase detector, a loop filter, a voltage-controlled oscillator (VCO), and a frequency divider in a feedback loop. The most common types of PLL are the digital PLL (DPLL), the linear PLL, the all-digital PLL (ADPLL), and the software PLL [2, 3]. The digital domain has many merits, including simplified calibration, increased predictability, and improved precision. As a result, analog circuits have been replaced by digital versions. PLLs allow high-performance microprocessors to have low power consumption. PLLs can be made process-independent and can be ported to a variety of technologies. Compared to analog and semi-digital PLLs, ADPLLs replace the VCO with a digitally controlled oscillator and a digital loop filter, resulting in higher performance, greater lock-in range, improved phase jitter performance, reliability, and speed and a reduction in size and cost [2].

A linear phase detector is one in which the output and phase errors have a linear transfer function. Phase frequency detectors (PFDs) measure phase variations between the input signal and a reference signal. The PFD generates two output signals (up and down). These signals are combined to form a single output signal, which is fed to the VCO as a control signal. This is accomplished with the help of a charge pump (CP) and loop filters. VCOs can be either current starved or source-couple-based. A current mirror and a current-starved inverter circuit are the two primary components of the current-starved VCO. The current-starved inverter ring oscillator is replicated by the current mirror circuit's control voltage [4].

Analog circuits are more difficult to miniaturize than digital circuits, and radio frequency (RF) signals are particularly sensitive to these modifications. As a result, digital circuits have only recently supplanted analog circuits. In terms of their manufacturing processes, digital circuits are significantly faster and easier to

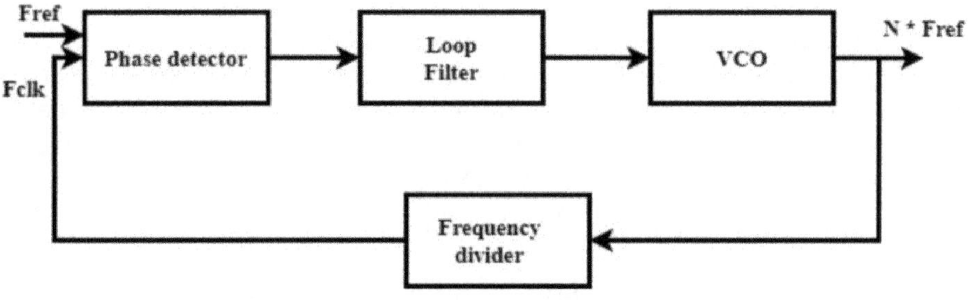

Figure 12.1. System-level block diagram of a PLL.

incorporate. Low-power technologies have emerged as a result of the increased demand for low-power devices. Various low-power approaches are utilized, including adiabatic logic [5], reversible logic, and adaptive voltage levels (AVLs). Adiabatic logic requires greater areas (transistor). AVLS is the preferred AVL. The AVLS circuit is made up of one pMOS and two nMOS transistors wired in parallel. The AVLS circuit is inserted into the supply voltage's source terminal.

This chapter describes a low-power modified PLL that can also operate at higher frequencies, making it appropriate for biomedical applications. The rest of the chapter is structured as follows: section 12.2 offers a literature review. Section 12.3 focuses on the design and implementation of the sub-blocks of a PLL and integrated PLL circuits in Cadence Virtuoso. Section 12.4 presents the results obtained and the inferences drawn. Section 12.5 presents the conclusions drawn from the findings as well as the future scope.

12.2 Literature review

The advantages of an ADPLL over a PLL with respect to programmability and stability were studied in [6]. The paper discusses the techniques used to design the individual units of an ADPLL, as well as differences between the various designs. Stability and programmability were used as the evaluation metrics. Reference [2] discusses the power usage, lock-in range, and frequency resolution of different permutations of individual units by examining the ADPLL's power usage, lock-in range, and frequency resolution. Prathibha *et al* [7] described a concept for an ADPLL that produced appropriate frequency signals using an accumulation-style digital controlled oscillator (DCO). A K-counter loop filter was utilized, and the ADPLL consumed 18.3 mW of power.

Madhusudan *et al* [8] designed and implemented a current-starved VCO (CSVCO) in Cadence. The frequencies available ranged from 53 MHz to 2.3 GHz and a power consumption of 848 μW was obtained. To decrease jitter, a D-type flip was employed. An ADPLL for wireless systems was developed by combining a master-slave D-type flip for time constant detector, a counter-based loop filter, and a three-stage ring oscillator DCO [9]. The ADPLL design described in [9] had a 50 ps locking time, a power usage of 26 mW, and a 4.7 GHz working frequency range. Reference [10] discusses the impacts of changing the bit-detection gap of a bang-bang phase detector to show the clock phases and phase updating interval; this study used MATLAB/Simulink models.

An ADPLL containing the phase fault of the phase detector as well as an up-down counter in a DPLL was provided as bits, according to Vaithianathan *et al* [11]. This phase fault was converted into an appropriate input for a DCO implemented in Simulink, and the simulated results were verified. Reference [12] discusses the use of an adaptive voltage level method with a D flip-flop to minimize power usage. The adaptive voltage level to ground (AVLG) approach consumed less power than an AVLS.

Dinesh *et al* [13] described ADPLL performance in terms of its parameters, including a mix of various elements, frequency resolution, jitter quality, power usage, and lock speed, among others. Different combinations of ADPLL components were used to achieve fine resolution and fast lock-in time; the resulting ADPLLs were appropriate for system-on-a-chip applications.

The DPLL described in [4] was developed in CMOS using the Cadence application and simulated using Hspice. This DPLL had a central frequency of 100 MHz, a lock range of 20 to 60 MHz, a locking time of 1.5 s, and phase noise of −105 dBc Hz^{-1}. A PFD DPLL and a quad DPLL were created by Devon *et al* [14]. Two DPLL designs were implemented; one had a 30 MHz CSVCO and the other design comprised a 400 MHz differential inverter VCO; the phase detector was implemented using an XOR gate and a fixed divider.

The application of PLLs was discussed in [15]. Reference [16] described a PLL that employed a low-voltage VCO, which was further modified using emitter-coupled logic (ECL) inverter cells. The lock range was improved by the inclusion of positive feedback resistance. The circuit was simulated in PSPICE using a minimum voltage of 2.5 V. Biomedical applications were also proposed for this PLL.

The power consumption increases in the presence of a dead zone (less than 20 ns), along with the locking time (340 ns), but the performance obtained is better. A dead-zone-free high locking time with a low-phase-noise PFD was proposed in [17]. The design of the PFD controlled the reference and feedback clock signals of the phase detector (PD) using the proposed NIKSTRO/SURAV latch-based sense amplifier. The design was realized using 45 nm CMOS technology.

ADPLL plays a role in the biomedical and communications fields. The power dissipation of DCOs can be minimized by designing them using XNOR gates and using delay elements to avoid direct contact between the supply lines [18]. In this study, metal–oxide–semiconductor (MOS) transistors were placed in a ring topology, and the design was implemented using Tanner Tools. The power consumption of the designed DCO was reduced by 35% with respect to the delay circuit.

Following a thorough review of the findings, different trends in phase detectors and PLL implementations were investigated. Power reduction solutions using CMOS technology were also investigated. In the following section, we discuss the design and realization of a low-power PFD and an integrated modified PLL.

12.3 Design and implementation of a modified PLL

This section discusses the implementation of the sub-blocks of the PLL and the integration of low-power sub-blocks to realize a PLL. Cadence computer-aided design (CAD) tools are used to realize the circuits and for simulation. The circuits are implemented on Virtuoso using 180 nm CMOS technology and simulated using Spectre.

12.3.1 Phase detector

The PFD shown in figure 12.2 is composed of two D flip-flops, whose inputs are always connected to logical 'high.' The two input signals which are fed to the phase detector consist of a frequency divider output, which is fed to the D clock, and the second signal, which is the external data signal. The 'up' signal is taken from the first

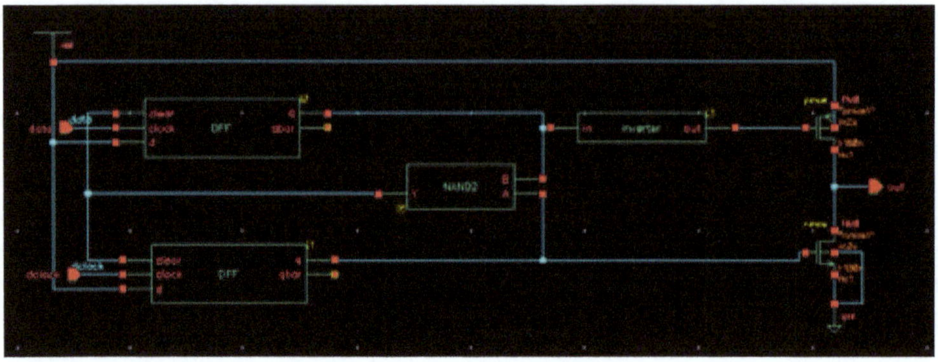

Figure 12.2. Phase frequency detector.

D flip-flop, whose input is the external data, and the 'down' signal is taken from the second D flip-flop, whose input is the D clock [11]. When the data signal is leading the D clock signal, the 'up' signal will be high until the PFD sees the rising edge of the D clock, and the 'down' signal will shift to and remain high until the rising edge of the data when the D clock is leading the data. When the PLL is in the locked condition, the phase error is zero. The difference between the phase of the input clock and the reference is represented by the equation (12.1):

$$\Delta\varnothing = \frac{\Delta t}{T_{\text{ref}}}, \tag{12.1}$$

where $\Delta\varnothing$ is the phase error, Δt is the time delay between the peaks of two input signals, and T_{ref} is the time period of the input reference signal.

The PFD output voltage, V_{PFD}, can be calculated using equation (12.2):

$$V_{\text{PFD}} = \frac{[\text{VDD} - 0]}{4_{*}\pi}{}_{*}\left[\Delta\varnothing\right]. \tag{12.2}$$

Thus, the gain of the PFD can be calculated using:

$$K_{\text{PFD}} = \frac{\text{VDD}}{4_{*}\pi}, \tag{12.3}$$

where K_{PFD} is the gain of the PFD (volts rad^{-1}).

A linear phase detector is shown in figure 12.3; its transfer function is linear between the output and the input phase error. The linear phase detector detects phase differences between signals belonging to a wide class of clock signal waveform types [9]. The block diagram of the proposed PLL system is shown in figure 12.4.

The phase-detector circuit can be a linear phase detector or a phase frequency detector. In the proposed architecture, the traditional loop filter that uses resistors and capacitors is replaced by a MOSFET-based loop filter. The VCO used is a

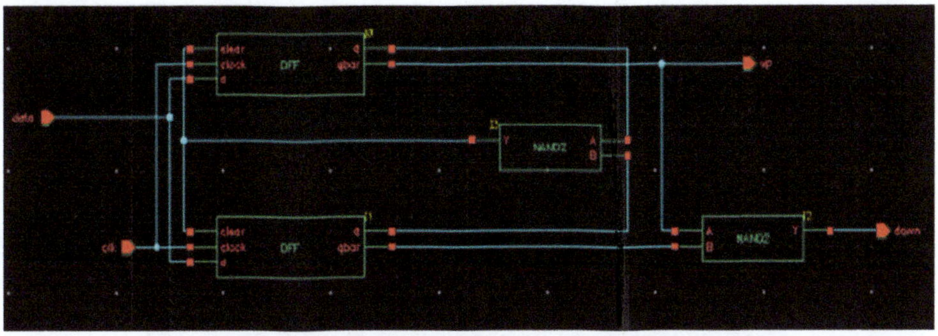

Figure 12.3. Schematic of a linear phase detector.

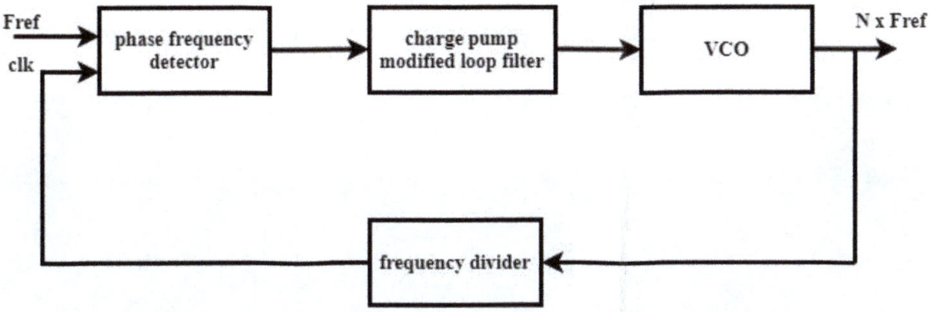

Figure 12.4. High-level block diagram of the proposed PLL architecture.

CSVCO. The frequency divider is realized using a synchronous counter, an XNOR gate, and a NAND gate.

12.3.2 Charge pump, loop filter, and current-starved VCO

The charge pump circuit and loop filter are shown in figure 12.5. The traditional resistor and capacitance (RC)-based loop filter is replaced by a MOSFET-based loop filter [19]. The charge pump circuit converts the two signals into one signal and feeds the result to the loop filter. The loop filter output drives the VCO. When the input signal lags behind the feedback clock, 'down' is high and the VCO output is low and vice-versa. The charge pump operation is discussed in detail in [20]. When both up and down signals are low, current flows to the loop filter. When up is low and down is high, the circuit discharges, and in the other cases, the output voltage is constant.

The CSVCO shown in figure 12.6 is composed of a current mirror circuit and current-starved inverters. The current-starved inverter is realized using a basic inverter circuit to limit the current drawn from the power supply. The current in the ring oscillator is mirrored using a current mirror. The CSVCO consists of five inverters. The current of the MOS is replicated in the inverters. The VCO output frequency is passed through a frequency divider before being fed to the PFD, as

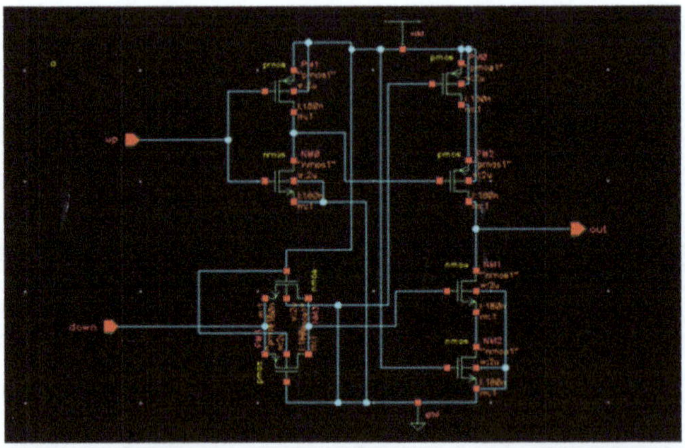

Figure 12.5. Charge pump and MOSFET-based loop filter.

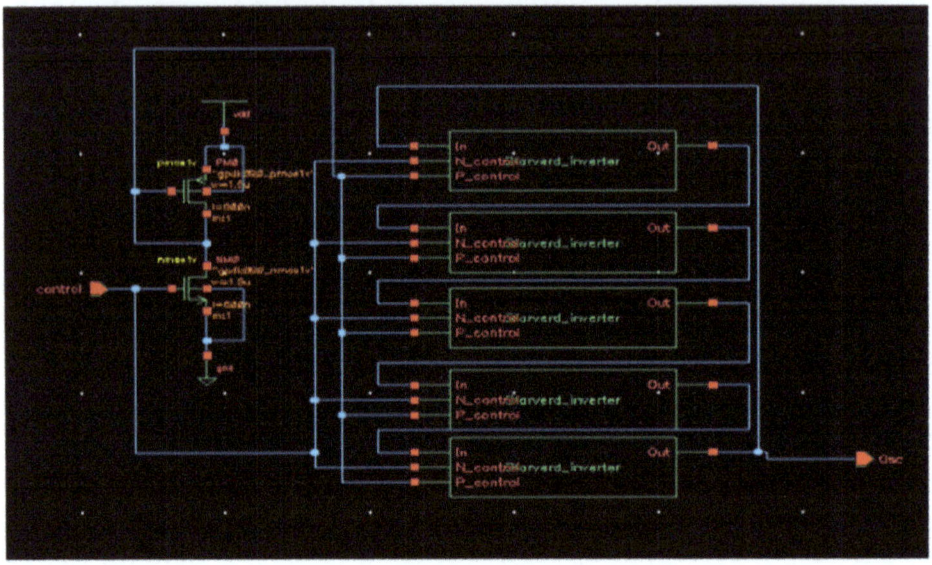

Figure 12.6. Current-starved VCO.

shown in figure 12.7. The 'divide by' counter is programmable and is realized using an array of XNOR and NAND gates and a synchronous counter.

12.3.3 AVLS approach

Adaptive voltage level techniques are preferred for low-power applications because AVLS circuits have tiny leakage currents. The basic AVLS circuit shown in figure 12.8 has one pMOSFET and two nMOSFETs connected in parallel. The pMOS of the circuit is fed by the clock input and the nMOS is connected to the drain terminal.

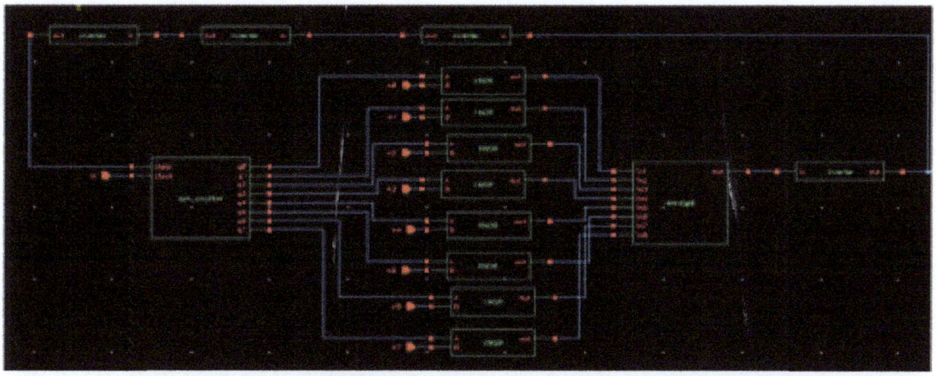

Figure 12.7. Frequency divider.

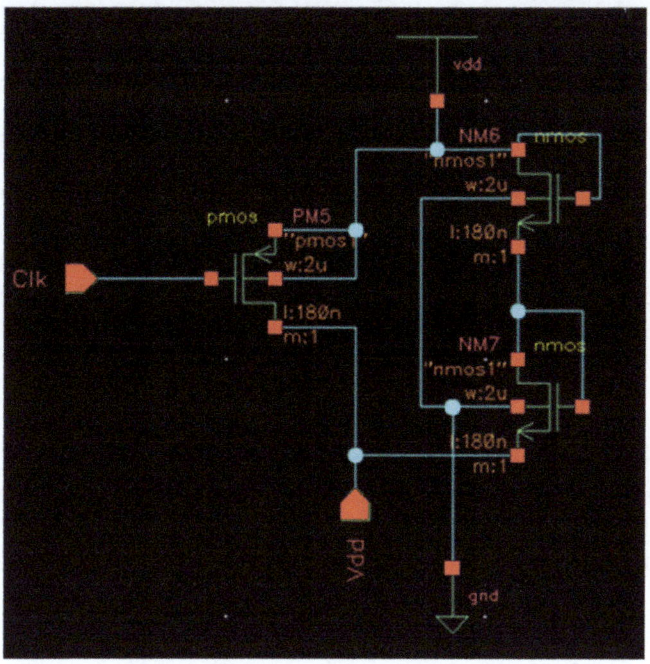

Figure 12.8. Schematic of an AVLS circuit.

12.3.4 Proposed modified PLL

The implementation details of various PLL architectures are discussed in this section.

A. PLL with a linear phase detector

The integrated architecture of a modified PLL which uses a linear phase detector and a charge pump circuit is represented in figure 12.9. The control voltage of the VCO is set to 1.8 V. The up and down signals from the linear PD are fed to the

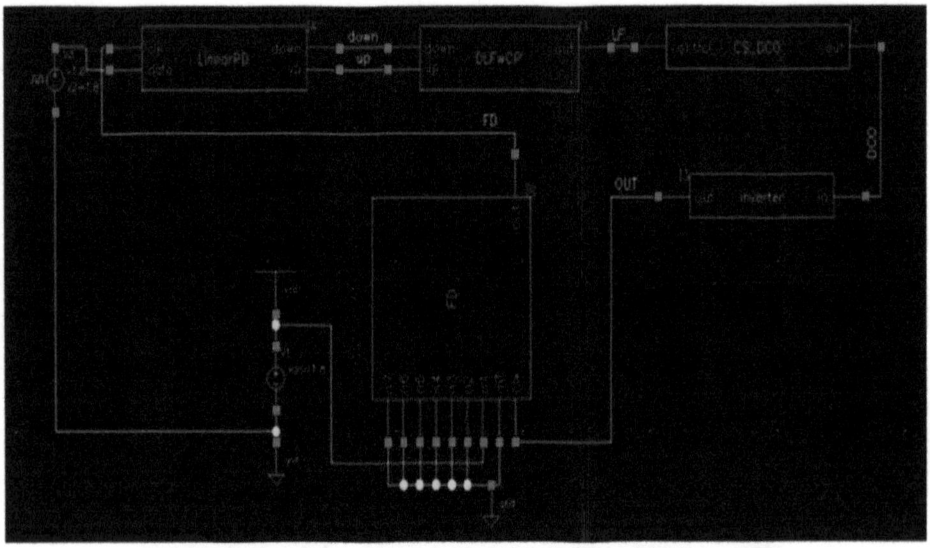

Figure 12.9. Modified PLL with a linear phase detector and an AVLS circuit.

digital charge pump circuit. The low-pass DLF frequency is fed to the DCO. The DCO output is fed to the frequency divider. The negative feedback from the frequency divider is fed to the linear phase detector. In the PLL circuit shown in figure 12.9, a dead-zone inner circuit is applied to the linear phase detector and the AVL technique is used for the control voltage to reduce the power consumption further.

B. PLL with a phase frequency detector

Figure 12.10 represents the integrated architecture composed of the sub-blocks explained in the previous section. The input data are fed to the PFD. The phase error output from the PFD is fed to the loop filter (LF). The low-pass LF frequency is fed to the VCO. The VCO is a five-stage ring oscillator circuit with a current-starved inverter circuit mirrored in all five stages. The VCO output is applied to the frequency divider.

12.4 Results and discussion

The simulations of the proposed PLL circuits were verified using Cadence Spectre. The output waveforms and simulation results for the PLL circuits are discussed in this section.

The output waveforms of the proposed PLL are shown in figures 12.11 and 12.12, respectively. The loop filter output is highly distorted, as shown in figure 12.11. The distortion at the output was the result of the applying the AVLS circuit to the loop filter using a charge pump. When the AVLS is applied to the PFD, VCO, and frequency divider, it is observed to cause much less distortion to the waveforms (as shown in figure 12.12), compared to figure 12.11. After we verified the simulated

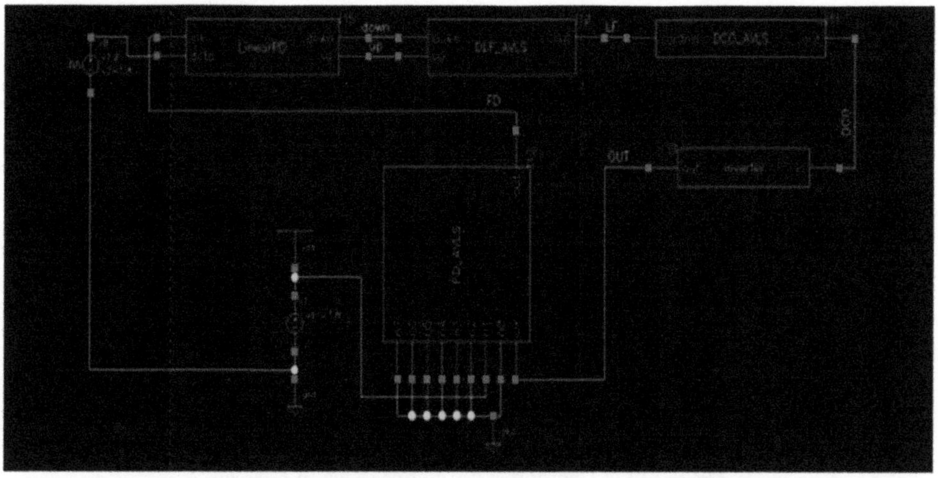

Figure 12.10. Modified PLL with PFD and AVLS circuits.

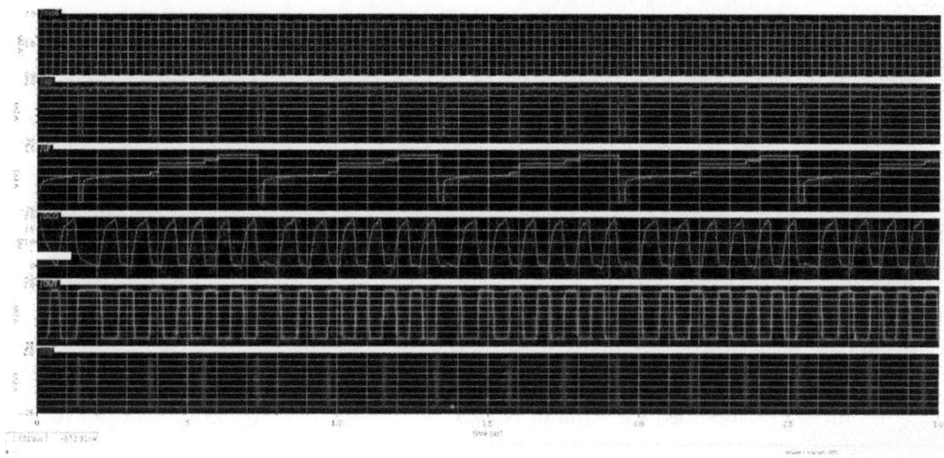

Figure 12.11. Distorted waveforms of the PLL circuit.

circuit results, we computed the power consumptions of the linear phase detector and phase frequency detector circuits.

The power consumption of the phase detectors discussed is displayed in table 12.1 for a frequency of 20 MHz. The power consumption and transistor count of the proposed PLL architecture realized in Cadence Virtuoso using 180 nm technology at 20 MHz are listed in table 12.2. There is a slight increase in the number of transistors due to the incorporation of AVLS, compared to other PLL architectures. The power consumption of the proposed PLL circuits is compared with the PLL circuit implemented in [21] at an operating frequency of 4 GHz. The values obtained are listed in table 12.3.

The Proposed-1 PLL is designed using a linear phase detector with an AVLS circuit and the Proposed-2 PLL is designed using a PFD with an AVLS circuit. For

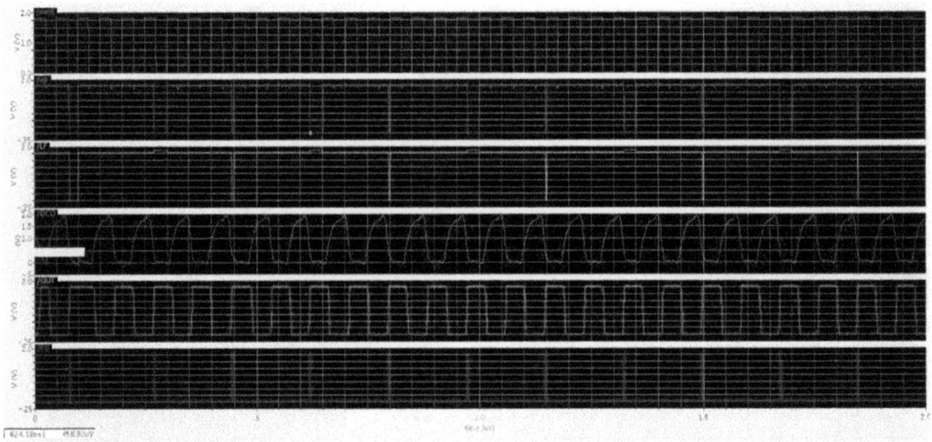

Figure 12.12. Undistorted waveforms of the PLL circuit.

Table 12.1. Power analysis of phase detectors.

S. no.	Phase detector architectures	Power (μW)
1	Linear phase detector	50.3
2	Phase frequency detector	66.2

Table 12.2. A comparison of PLL architectures w.r.t. power and transistor count at a frequency of 20 MHz.

S. no.	PLL architecture	Power (mW)	No. of transistors
1	Linear PD, modified LF without AVLS	0.9551	616
2	Linear PD, modified LF with AVLS	0.3713	642
3	Linear PD, AVLS to all subcircuits except CP with LF	0.3439	639
4	PFD, AVLS to all subcircuits	0.3168	642
5	PFD, AVLS to all subcircuits except CP with LF	0.2318	639

Table 12.3. Comparison of PLL architectures w.r.t. power at a frequency of 4 GHz.

S. no.	PLL architectures	Power (mW)
1	PLL in [21]	1.81
2	Proposed-1 (linear PD-PLL)	1.3
3	Proposed-2 (PFD PLL)	0.6

both circuits, a MOSFET-based loop filter is used instead of an RC circuit. It can be seen from table 12.3 that the Proposed-1 PLL obtains a power reduction of 0.5 mW and the Proposed-2 PLL obtains a power reduction of 1.21 mW. Hence, the proposed PLL circuits are power efficient and can be used for biomedical applications that require circuits with very minimal power consumption.

With the rise of chronic disease and ailments related to age, there is a continuous demand for efficient medical electronic devices, which are required for the measurement, monitoring, and recording of body functions. The essential requirements of any biomedical implantable device are low power, low noise, and lifetime performance. The power consumption of the proposed PLLs is in the mW range, and the number of transistors is also reduced compared to conventional PLLs; therefore, the proposed PLLs can be used in technical biomedical implantable transceivers that require very low power consumption and work in high-frequency ranges for precise diagnoses.

12.5 Conclusions

With the demand for reduced power consumption in the medical field, there is a gradual shift from the analog domain toward the digital domain. To design a power-efficient PLL for biomedical devices, it is necessary to minimize the power in all the subcomponents of the PLL. The Proposed-1 PLL, which has a linear phase detector, a charge pump with a modified loop filter, a CSVCO, and a frequency divider with the AVLS technique, has a power reduction of 28%, and the Proposed-2 PFD PLL has a power reduction of 66%. The proposed PLL circuits operate at high frequencies and consume very small amounts of power; they are therefore preferable for biomedical applications. The PLL circuits can be further optimized and various other power reduction techniques can be implemented to reduce the power consumption again. The designed PLL can be integrated into biomedical devices to make them power efficient.

Bibliography

[1] Zvyagin A V, FitzGerald J B, Silva K K M B D and Sampson D D Real-time detection technique for doppler optical coherence tomography *Opt. Lett.* **25** 1645–47

[2] Lata K and Kumar M 2013 ALL digital phase-locked loop (ADPLL): a survey *Int. J. Future Comp. Commun.* **2** 551–5

[3] Bency M, Prasanna S R, Shreya G P and Premananda B S 2021 PFD with dead zone based low power modified phase lock loop using AVLS technique *6th Int. Conf. on Communication and Electronics Systems (ICCES)* pp 212–17

[4] Fernandez D and Manandhar S 2003 *Digital phase locked loop* University of Maine pp 1–59 https://ece.umaine.edu/research/vlsi/student-integrated-circuit-designs/fernandez-and-manandhar/

[5] Ganavi M G and Premananda B S 2019 Design of low-power square root carry select adder and wallace tree multiplier using adiabatic logic *Emerging Research in Electronics, Computer Science and Technology (Lecture Notes in Electrical Engineering* vol 545) (Berlin: Springer) pp 767–81

[6] Singhal A, Madhu C and Kumar V 2014 Designs of all digital phase locked loop *2014 Proc. of Recent Advances in Engineering and Computational Sciences (RAECS)* pp 1–5 UIET Punjab University Chandigarh
[7] Kavyashree P and Reddy V S 2018 Implementation of ALL digital phase locked loop *Int. J. Res. Anal. Rev.* **5** 1386–9
[8] Madhusudan K and Kalmeshwar N H 2013 Design of linear and wide range current starved voltage controlled oscillator for PLL *Int. J. Cybern. Inform.* **2** 23–30
[9] Swetha R J, Manjula A and Ruhan B 2014 Design of ALL digital phase locked loop for wireless applications *Int. J. Eng. Technol.* **7** 836–40
[10] Ahmed S I and Kwaśniewski T A 2005 An ALL-digital data recovery circuit optimization using matlab/simulink *2005 IEEE Int. Symp. on Circuits and Systems* pp 4485–8
[11] Vaithianathan V, Eshwar V H, Kamalakannan S and De Raja J 2017 ALL-digital phase locked loop with an up-down counter using simulink *2017 Int. Conf. on Wireless Communications, Signal Processing and Networking (WiSPNET)* pp 2424–7
[12] Malipatil S 2015 Design of a low power D-flip flop using AVL technique *Int. J. Adv. Res. Comput. Commun. Eng.* **4** 291–3
[13] Dinesh R and Marimuthu R 2016 A survey on ADPLL components and their effects upon power, frequency and resolution *Int. J. Appl. Eng. Res.* **11** 1569–74
[14] Wang K, Tang Z, Ge T and Li H 2011 Design and simulation of a VCO in CMOS digital PLL *2011 4th Int. Conf. on Intelligent Computation Technology and Automation* pp 749–52
[15] Nargund A A, Abhilash Kashyap C M, Prashanth D, Premananda B S and Sharma A 2016 A novel method of wideband acquisition and anti-sideband lock in PM receivers using FFT *2016 IEEE Int. Conf. on Recent Trends in Electronics, Information & Communication Technology (RTEICT)* pp 2098–101
[16] Chakraborty B and Pal R R 2007 Design of a PLL with improved locking range and its biomedical applications *Int. J. Electron.* **94** 671–82
[17] Gunda S and Ernest Ravindran R S 2020 A novel low power, minimal dead zone digital PFD for biomedical applications *Int. J. Adv. Comput. Sci. Appl.* **11** 333–43
[18] Gunda S and Ernest Ravindran R S 2019 Design and simulation of low power consuming digital controlled oscillator in all digital phase locked loop *Int. J. Innov. Technol. Explor. Eng.* **8** 3801–6
[19] Naik M D K 2016 *Design and Analysis of Efficient Phase Locked Loop* National Institute of Technology, Rourkela
[20] Patel N D and Naik A P 2018 A low jitter—low phase noise wideband digital phase-locked loop in nanometer CMOS technology *Int. J. Electron. Commun. Eng. Technol.* **9** 1–12
[21] Sen Y and Jain N 2014 Design and implementation of phase locked loop using current starved voltage controlled oscillator *J. Adv. Electron. Electric Eng.* **4** 637–44

Chapter 13

Image multiplication with a power-efficient approximate multiplier using a 4:2 compressor

Lavanya Maddisetti, Ranjan K Senapati and J V R Ravindra

13.1 Introduction

Multiplication is the most fundamental and widespread operation in almost all applications, such as digital signal processing, image processing, and neural network computation. However, power consumption is increased and a large silicon area is occupied by the partial product reduction stage of multiplication operations. Hence, approximate computing has been developed to reduce the power consumption. In this chapter, we describe the creation of an 8 × 8 unsigned Dadda multiplier that includes a proposed 4:2 compressor and uses a probabilistic pruning type of approximation technique. This Dadda multiplier is employed for pixel-by-pixel multiplication of two images. According to the simulation results obtained from Cadence Virtuoso, it can be seen that the power consumption is decreased by 82.83% compared to that of an exact multiplier.

It is critical to increase the power efficiency of the digital circuits used in current chip-based platforms. Machine learning, image identification, object detection, and hardware neural networks are a few of the relevant applications that are error-resistant. Approximate computation is an important way to achieve performance gains in terms of power, speed, and area for these types of application. Arithmetic blocks, such as compressors and multipliers, are among the most power-intensive digital blocks. Therefore, hardware-level approximations for these blocks are studied in this chapter. The proposed 4:2 compressor is placed in a Dadda multiplier, which is used in an image processing application, i.e. multiplication. Arithmetic circuits such as adders and multipliers play significant roles in microprocessors and digital signal processors, which use multipliers to implement DSP algorithms such as filtering and convolution, and they are also used in logic units [1]. Addition and multiplication are the most frequently used operations in very large

doi:10.1088/978-0-7503-3919-3ch13

scale integration (VLSI) systems. The steps involved in the multiplication operation are partial product generation, partial product reduction (PPR), and final addition with a carry.

Among these three, the partial product reduction stage consumes large amounts of power and silicon area and increases delay, as it consists of repeated addition, which involves different adder circuits. Thus, to reduce the above parameters, compressors have become critical components that have replaced all the adder circuits of a multiplier [2]. By convention, compressors are structured as 3:2, 4:2, 5:2, or higher orders. They all consist of XOR–XNOR gates and transmission-gate-based multiplexers. It is shown in [2] that a 4:2 compressor can be designed that uses low-power combined XOR–XNOR gate-based full-adder circuits, from which the inverters are removed, with a resulting reduction in transistor count from 40 to 28. Various low-power higher-order compressors, such as 4:2 and 5:2 compressor modules, are examined in [3]. This paper also compared the delay and power consumption of various circuit designs for 4:2 compressors. The authors of [4, 5] presented unique architectures for high-speed, low-power 3:2, 4:2, and 5:2 compressors that can operate at ultra-low voltages. Reference [6] presents decimal compressors that can handle decimal multipliers.

The authors of [7] explored a broad range of techniques that trade off accuracy for improved efficiency or energy consumption in approximate computing. For any given specification, if an approximation is applied to the exact circuit, the accuracy changes a little, resulting in a reduction to the delay, power, and circuit size. This type of approximate theory and its applications are described in [8]. New designs for approximate arithmetic circuits, such as adders, multipliers, and dividers, are reviewed and classified in [9]. The authors of [10] presented a survey of approximate computation techniques along with strategies for finding approximate circuit parts and monitoring performance efficiency. The authors of [11] proposed a novel approach for realizing inexact circuits, called probabilistic pruning. This method relies on architectural changes that prune the portions of circuits that are less active, resulting in considerable energy, delay, and area savings. The authors of [12] presented a new method known as Inexact Logic Minimization that depends on synthesizing an inexact circuit, resulting in no hardware overhead.

Gupta *et al* [13] proposed several inaccurate or approximate full-adder cells with decreased transistor complexity to take advantage of numerical accuracy relaxation, demonstrating a logic complexity reduction method. The construction of imprecise bio-inspired computational blocks was described in [14]; they provide approximate results and are more effective in terms of area, speed, and power consumption than their precise counterparts. An inexact N-bit ripple carry adder architecture was examined in [15]. The aim of [16] was to introduce a new design method for approximating multipliers by including various probability terms and modifying the partial products of the multiplier. Jiang *et al* [17] focused on existing approximate adder designs and compared them in terms of both error and circuit characteristics. Error analyses were introduced in [18] to measure the efficiency and power efficiency of approximated probabilistic adders. For the compressor and multiplier presented in this chapter, the error metrics reported in [19] are considered.

An approximate 4:2 compressor [20] was placed in a Dadda multiplier which was used in an infinite impulse response filter to monitor the speed of the output signal. Maddisetti *et al* [21] introduced an approximate technique called inexact logic minimization and implemented it on a 4:2 compressor that minimizes power and area by changing a few output bits. An approximate multiplier architecture that uses both truncation and compressor approximation was outlined in [22]. The authors of [23] demonstrated an approximate compression technique for biosignals in a wearable healthcare tracking device. Various other 4:2 compressors were exhibited in [24], which were able to shift from an exact mode to an approximate mode and vice versa. Reference [25] described a novel design that incorporated an error-recovery module based the modification of a previous approximate 4:2 compressor design. A novel approximate 4:2 compressor and its circuit implementation were proposed in [26] for error-resilient multiplication using a low supply voltage. The taxonomy of previous research work is classified and illustrated in figure 13.1.

The analysis and construction of two recent approximate 4:2 compressors, which can be used in a multiplier, were discussed in [27]. The authors of [28] proposed a new architecture for compressor-based approximate multipliers that used recursive multiplication to manipulate the partitions of partial products. To demonstrate the efficacy of the suggested designs, an image processing application was also presented. With regard to design, two of the proposed multiplier architectures have outstanding capabilities for image multiplication. Design methods for approximate 4:2 and 5:2 compressors were suggested in [29] to reduce the partial product stages in multiplication. A new imprecise 4:2 compressor that can be used in image processing multipliers was presented in [30]. Two approximate 4:2 compressors were proposed in [31] for incorporation into a Dadda multiplier, which was further used in image smoothing and multiplication applications. A revised Dadda multiplier

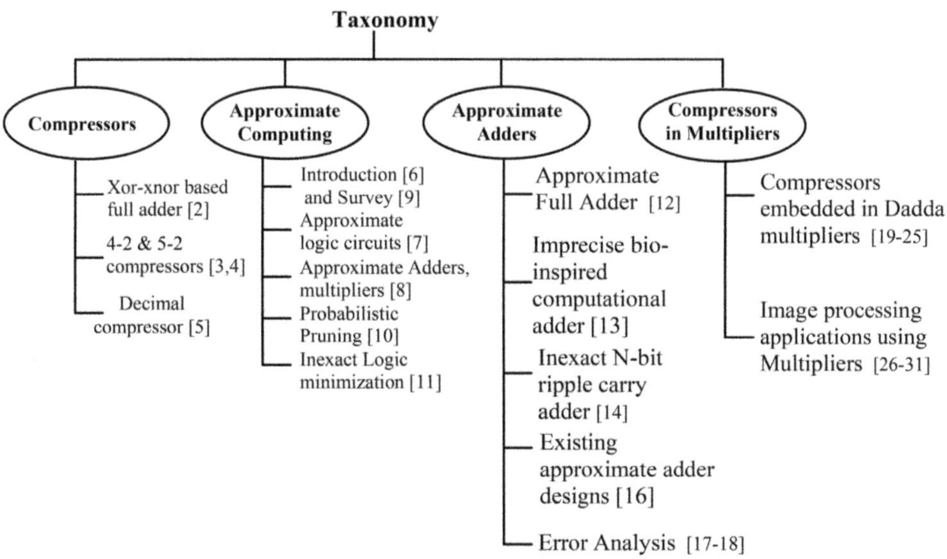

Figure 13.1. Taxonomy of related work.

architecture was presented in [32] to make efficient use of the proposed compressor and reduce output error. Several image processing applications have been used to evaluate multiplier efficacy. Strollo *et al* [33] presented a survey and a comparison of approximate 4:2 compressors. They also introduced a novel approximate compressor, bringing the total number of approximate 4:2 compressors to twelve. The circuits analyzed were used to create different orders of multiplier. Ravindra *et al* [33], and Ritajit *et al* [34] proposed similar decoders.

13.2 Existing 4:2 compressor

The compressor is an essential block of a multiplier circuit; it reduces partial products. The full adder is a basic compressor, which was introduced as 3:2 compressor with three inputs compressed to two outputs. The subsequent prominent compressor was the 4:2 compressor, which has five inputs: X_1, X_2, X_3, X_4, and C_{in} and three outputs: Sum, Carry, and C_{out}. Figure 13.2 depicts the basic structure of a 4:2 compressor.

C_{in} accepts the output of the preceding lower-order compressor, and C_{out} is an output sent to the subsequent higher-order compressor. The equations that define the outputs of the 4:2 compressor are as follows:

$$\text{Sum} = X_1 \oplus X_2 \oplus X_3 \oplus X_4 \oplus C_{in} \tag{13.1}$$

$$\text{Carry} = X_4(X_1 \oplus X_2 \oplus X_3 \oplus X_4)' + C_{in}(X_1 \oplus X_2 \oplus X_3 \oplus X_4) \tag{13.2}$$

$$C_{out} = X_3(X_1 \oplus X_2) + X_1(X_1 \oplus X_2)'. \tag{13.3}$$

The proofs of the above equations are derived in appendix A. Based on these equations, many researchers have introduced distinct 4:2 architectures with prominent changes. In this chapter, an existing 4:2 compressor is examined along with the proposed approximate 4:2 compressor. The sum expression of an exact 4:2 compressor in terms of two input XOR–XNOR gates is as follows:

$$\begin{aligned}\text{Sum} = {} & C_{in}[(X_1 \oplus X_2)(X_3 \oplus X_4)' + (X_1 \oplus X_2)'(X_3 \oplus X_4)]' \\ & + C_{in}'[(X_1 \oplus X_2)(X_3 \oplus X_4)' + (X_1 \oplus X_2)'(X_3 \oplus X_4)].\end{aligned} \tag{13.4}$$

The precise block diagram of a 4:2 compressor is shown in figure 13.3 for the above sum and carry equations.

The truth table of a regular 4:2 compressor is given in [26]. A six transistor (T) pass-transistor-logic-based XOR–XNOR circuit has been employed, in which there is no reduction in the voltage swing; the two-input multiplexer (MUX) is based on

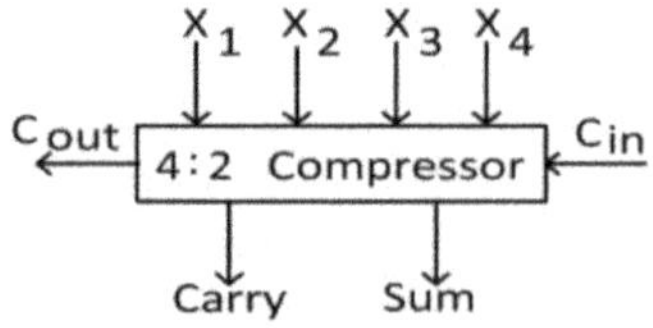

Figure 13.2. Structure of a compressor with four inputs and two outputs. © 2007 IEEE. Reprinted, with permission, from [4].

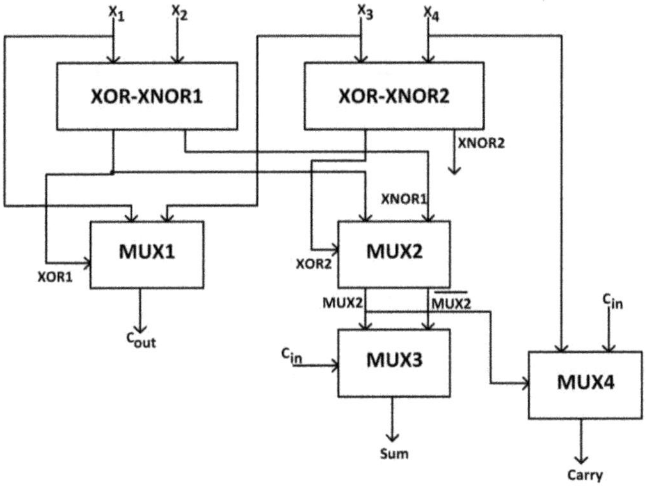

Figure 13.3. XOR–XNOR-based 4:2 compressor. © 2007 IEEE. Reprinted, with permission, from [4].

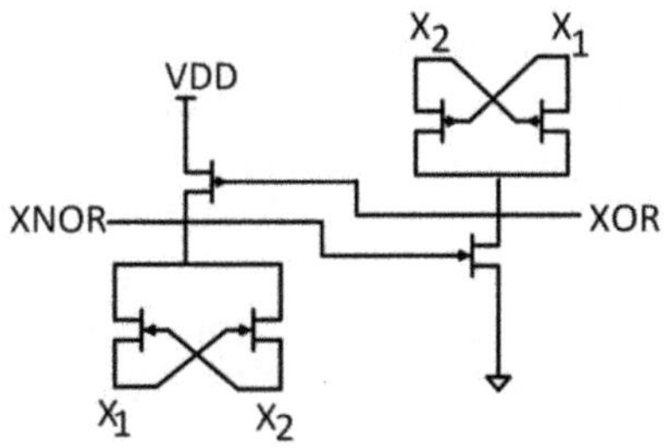

Figure 13.4. 6T XOR–XNOR. © 2004 IEEE. Reprinted, with permission, from [5].

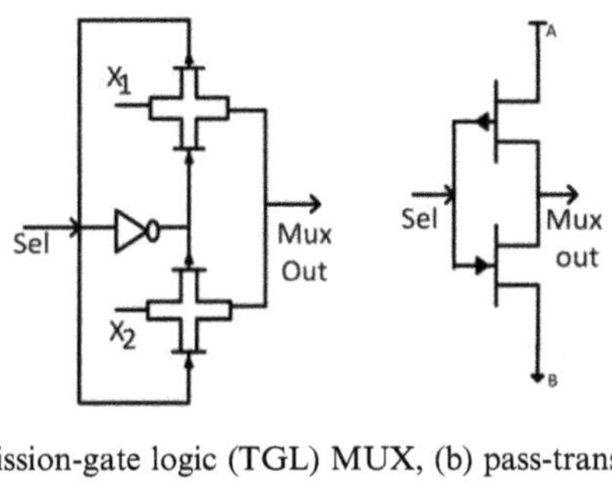

Figure 13.5. (a) Transmission-gate logic (TGL) MUX, (b) pass-transistor logic (PTL) MUX.

transmission-gate logic. The elementary units of the 4:2 compressor, the XOR–XNOR and the MUX, are represented in figures 13.4 and 13.5(a), respectively.

In a regular 4:2 compressor, two input multiplexers with pass-transistor logic are used in a transmission-gate logic (TGL)-based multiplexer, which shrinks the power and delay compared to that of a 4:2 compressor with a TGL multiplexer. The pass-transistor logic (PTL)-based multiplexer, which has a minimal number of transistors, is portrayed in figure 13.5(b).

13.3 Proposed work

This section describes the proposed approximate 4:2 compressor as well as a multiplier that embeds the proposed compressor. The probabilistic pruning kind of approximation strategy [11, 34] is used to construct the 4:2 compressor approach shown here. In this method, the multiplexer at the Sum output of the exact 4:2 compressor depicted in figure 13.3 (section 13.2) is pruned.

MUX3 is discarded, converting the MUX2 output terminal into the Sum output without affecting the C_{out} and Carry expressions. The figure 13.6 shows the approximate, novel 4:2 compressor; the pruned blocks of the compressor are depicted in red.

The novel 4:2 compressor is constructed using 15 transistors. XOR-XNOR1 requires 6T and XOR2 requires 3T and 2T in each MUX. The exact Sum equation (13.4) is reduced to the following approximated Sum_A expression, while equations (13.2)–(13.3) for C_{out} and Carry remain the same as for the exact calculation.

$$Sum_A = (X_3 \oplus X_4)' (X_1 \oplus X_2) + (X_3 \oplus X_4)(X_1 \oplus X_2)'.$$

Figure 13.6 exhibits a circuit with the fewest errors possible. The Sum output has eight errors, which result in a 25% error rate, and C_{out} and Carry have no errors. The power consumption of a Dadda multiplier is reduced by embedding the novel 4:2 compressor in it. Figure 13.7 shows the architecture of the 8 × 8 Dadda multiplier with the proposed 4:2 compressor in the PPR stage.

Figure 13.7 shows the partial product generation (first) stage built using AND gates. The AND gates are arranged in the form of eight 4:2 compressors, two full adders, and two half adders [26]. In the PPR (second) stage, ten proposed 4:2

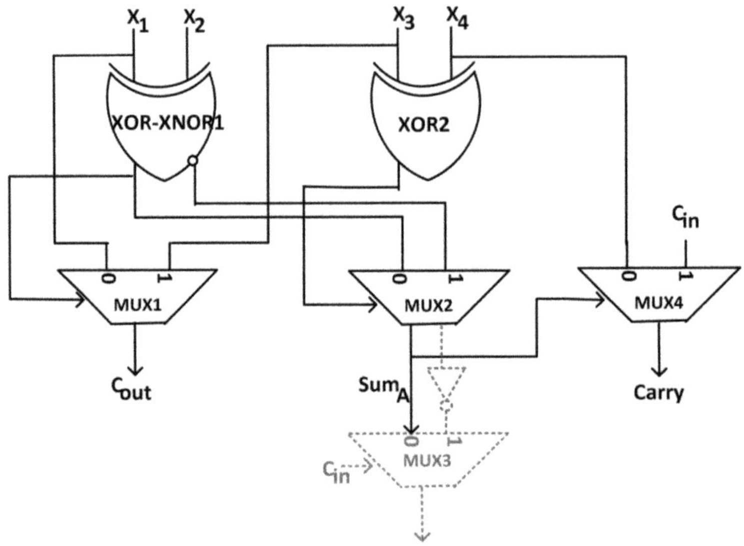

Figure 13.6. Proposed approximate 4:2 compressor.

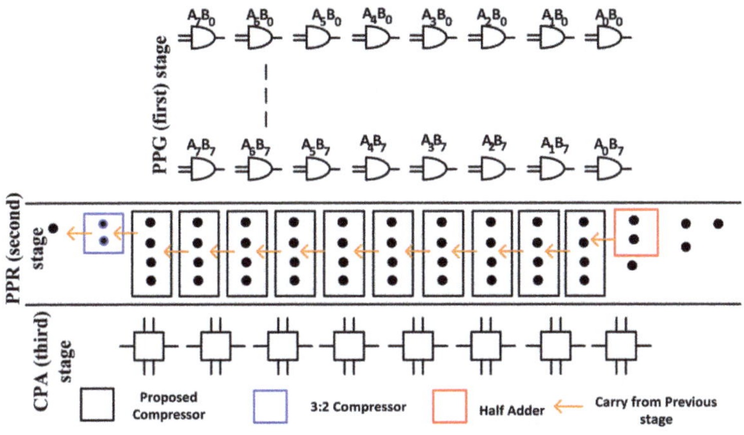

Figure 13.7. Novel 4:2 compressor in a Dadda multiplier.

Table 13.1. Power, area, and transistor count of the 4:2 compressor.

Compressor	No. of transistors	Power (nW)	Area (μm^2)
Exact [4]	50	178.9	29.893
Venkatachalam [16]	40	157.4	23.522
VLSID [21]	34	110	20.34
ICM [20]	30	105.2	16.989
Momeni [27]	28	102.4	16.728
CAEE [19]	19	94.53	14.36
Proposed	15	83.23	9.641

compressors, one full adder, and one half adder are used; finally, a carry-propagate adder with eight full adders is used in the third stage of the multiplier.

13.4 Results and discussion

The 4:2 compressors and multipliers were simulated using Cadence Virtuoso's Spectre simulator. Layout XL from Cadence was used in the 45 nm technology node to determine the area, and error analysis of the compressors and multipliers was performed using MATLAB.

13.4.1 Simulations of 4:2 compressors and multipliers

The proposed approximate 4:2 compressor is compared with existing circuits in terms of transistor count, power, and area. The 4:2 compressor simulation results are displayed in table 13.1.

Compared to the exact circuit, the proposed compressor improves the power consumption and area by 53.47% and 67.74%, respectively. Table 13.1 shows that the power consumption and area of the 4:2 compressors differ, depending on the number of transistors used. The graph in figure 13.8 illustrates how these parameters change as the number of transistors increases.

Each of the compressor circuits tabulated in table 13.1 is placed in the PPR stage of a Dadda multiplier, as shown in figure 13.7. Simulations of all these multipliers were performed; the performance metrics obtained are listed in table 13.2.

The first and third stages of these multipliers are exact designs (the same designs are used in all five multipliers), while the compressors embedded in the PPR stage of the multipliers differ according to table 13.1. As a result of the variations in the compressor circuits, the area and power values change, as depicted in figure 13.9.

13.4.2 Error analysis

There is a variation between the obtained and exact outputs when approximations are implemented in a formerly exact compressor. To observe the differences in these outputs, the error metrics used in [19] are utilized for 4:2 compressors and

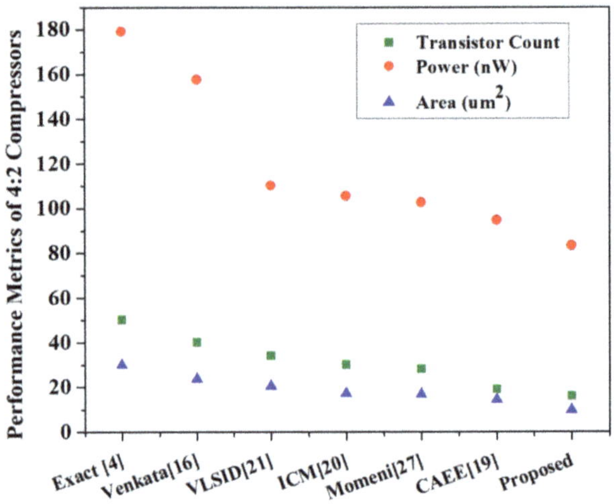

Figure 13.8. Simulation analysis for exact and approximate 4:2 compressors.

Table 13.2. Exact and approximate multiplier parameters.

Dadda multiplier using 4:2 compressor	Power (μW)	Area (μm^2)
Exact [4]	124.33	6108.9327
Venkatachalam [16]	97.02	5692.7025
VLSID [21]	38.76	5417.696
ICM [20]	32.59	5398.725
Momeni [27]	27.24	5194.8056
CAEE [19]	23.65	4936.46
Proposed	21.37	4647.8306

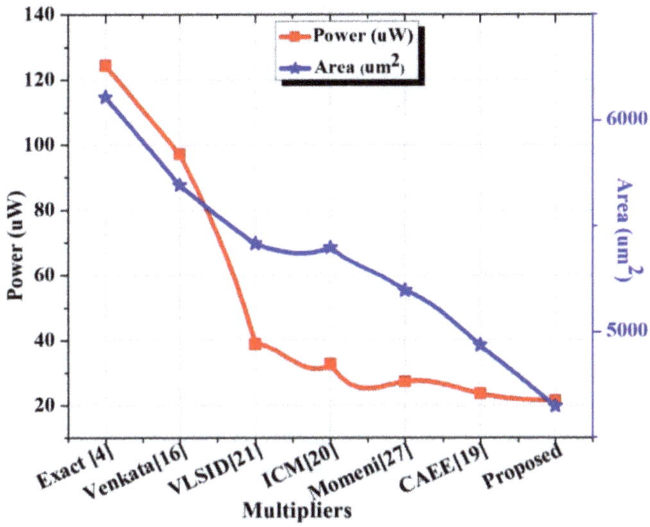

Figure 13.9. Power and area of exact and approximate multipliers.

Table 13.3. Error analysis for 4:2 compressors.

4:2 compressor	Errors	ER (%)	MED	NMED
Exact [4]	0	0	0	0
Venkatachalam [16]	5	31.25	3.3432×10^6	1.044×10^5
VLSID [21]	8	25	5.535×10^7	1.729×10^6
ICM [20]	10	31.25	4.394×10^7	1.959×10^6
Momeni [27]	12	37.5	7.551×10^7	2.359×10^6
CAEE [19]	8	25	3.412×10^7	1.0662×10^6
Proposed	8	25	1.203×10^3	37.6025

multipliers. The number of errors present, the corresponding error rate (ER) in the sum, the mean error distance (MED), and the normalized mean error distance (NMED) of the selected 4:2 compressors are given in table 13.3.

The proposed circuit has a 25% error rate when the C_{out}, Carry, and Sum outputs are considered, but the novel approximation 4:2 compressor and those described in [19, 21] have the same error rate, since those papers only considered the Sum output in the error rate. The design of [16] only had four (X_4, X_3, X_2, and X_1) inputs and two outputs (Carry and Sum). As a result, instead of 32 inputs, error analysis was performed for 16 inputs, yielding a 31.25% error rate for [16]. A graph of the number of errors and the error rates of the approximate compressors is shown in figure 13.10.

The error rate varies depends on the number of errors present in each approximate 4:2 compressor. When the same error analysis is performed on the different multipliers shown in table 13.2, the error metrics [19] shown in table 13.4 are obtained.

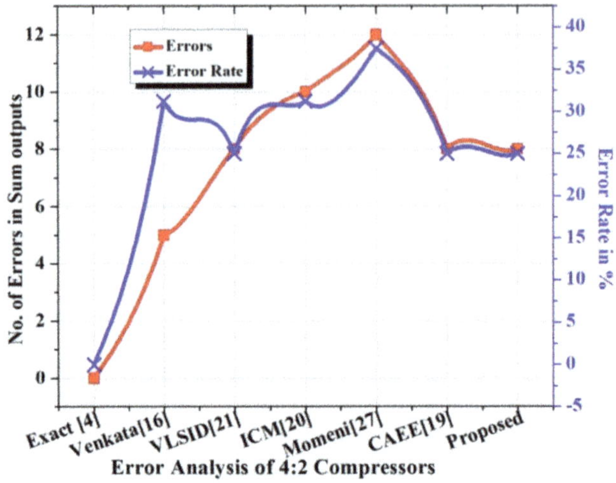

Figure 13.10. Error rates of the 4:2 compressors.

Table 13.4. Error analysis of multipliers.

Multipliers with proposed compressor	MED	NMED ($\times 10^{-7}$)
Exact [4]	0	0
Venkatachalam [16]	0.1161	18.541
VLSID [21]	0.1428	20.196
ICM [20]	5.1816	7.9686
Momeni [27]	1.928	28.122
CAEE [19]	0.0195	3.0037
Proposed	0.0584	15.674

13.4.3 Image multiplication

This section of the chapter discusses how an approximate multiplier can be used to multiply two images. In image multiplication, two input images are multiplied using their corresponding pixel values to produce an output image. Table 13.5 shows the multiplication of two images in steps which are implemented using Python code.

As an aid to visual understanding, we consider two images, Sailboat and House. The multiplications of the Sailboat and House images using the CAEE [19], VLSID [21], Momeni [27] and proposed approximate multipliers are shown in figure 13.11.

The image quality is estimated using the peak signal-to-noise ratio (PSNR) based on the mean square error (MSE). The definitions of MSE and PSNR [30] are given by equations (13.5) and (13.6):

$$\mathrm{MSE} = \frac{1}{\mathrm{MN}} \sum_{i=0}^{M-1} \sum_{j=0}^{N-1} [\mathrm{f}(x, y) - \mathrm{f}'(x, y)]^2, \tag{13.5}$$

Table 13.5. The procedure for image multiplication.

Step 1.	Consider two images, imagel and image2; each one has a size of 512 × 512
Step 2.	Extract the pixels of imagel and image2 and store them in matrix1 and matrix2, respectively
Step 3.	Convert matrix1 and matrix2 into list1 and list2
Step 4.	Convert all the pixels in list1 and list2 to their binary forms
Step 5.	Multiply the corresponding binary pixels of list1 and list2 using the existing and proposed approximate multipliers
Step 6.	Convert the the resultant multiplied output lists to the 'int' type and divide them by 256
Step 7.	Convert the 'int' output list into matrix form
Step 8.	Convert the output matrix into an image

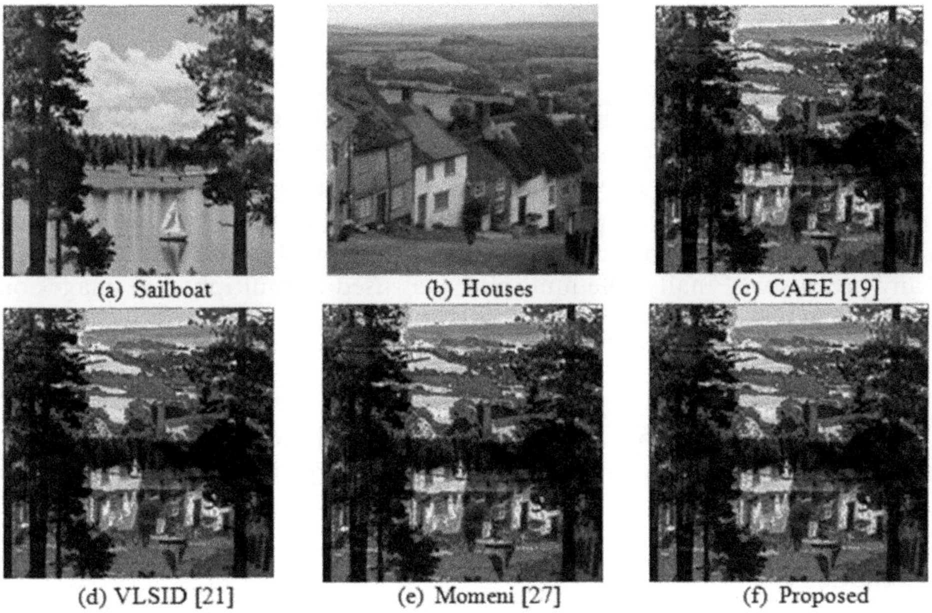

Figure 13.11. Multiplication of the Sailboat and House images performed by existing and proposed approximate compressor architectures using unsigned 8 × 8 Dadda multipliers.

$$\text{PSNR} = 10\log_{10}\frac{255^2}{\text{MSE}}. \tag{13.6}$$

M and N in equation (13.5) are the number of rows and columns of an image matrix; $f(x,y)$ and $f'(x,y)$ represent exact and approximate pixel values, respectively. In equation (13.6), 255 indicates the maximum pixel value in the image. Table 13.6 displays the PSNR values obtained for image multiplication performed by the approximate multipliers.

Table 13.6. PSNR values of images using the approximate multipliers.

Approximate multiplier	PSNR (dB)
VLSID [21]	24.33
Momeni [27]	25.56
CAEE [19]	27.16
Proposed	27.88

Python code is used to determine the PSNR values. Compared to the other three images, the image produced by the proposed multiplier design has better image quality and consumes a small amount of power.

13.5 Conclusions

In this chapter, we have described and analyzed the application of the approximate computing technique to digital circuits, such as 4:2 compressors, to achieve low power consumption. Rather than voltage scaling or other approximation methods, we used a probabilistic pruning type of approximation. The proposed compressor was placed in the partial product generation stage of an unsigned 8 × 8 Dadda multiplier, which is the most power-consuming block of any processor. Finally, the multiplier was used to multiply two images on a pixel-by-pixel basis.

Acknowledgements

This research was carried out at C-ACRL, Vardhaman College of Engineering. The authors would like to express their gratitude to the organization and staff for their unconditional support.

Appendix A

The Sum, Carry, and C_{out} expressions of an exact 4:2 compressor are derived as follows:

A.1 Expressions for the Sum

$$\text{Sum} = X_1 \oplus X_2 \oplus X_3 \oplus X_4 \oplus C_{in}$$

$$\text{Sum} = [(X_1 \oplus X_2) \oplus (X_3 \oplus X_4) \oplus C_{in}]$$

$$\text{Sum} = \left[(X_1 \oplus X_2)(X_3 \oplus X_4)' + (X_1 \oplus X_2)'(X_3 \oplus X_4)\right] \oplus C_{in}$$

$$\begin{aligned}\text{Sum} = {} & C_{in}[(X_1 \oplus X_2)(X_3 \oplus X_4)' + (X_1 \oplus X_2)'(X_3 \oplus X_4)]' \\ & + C_{in}'[(X_1 \oplus X_2)(X_3 \oplus X_4)' + (X_1 \oplus X_2)'(X_3 \oplus X_4)]\end{aligned}$$

A.2 Carry expressions

$$\text{Let, } W = X_1 \oplus X_2 \oplus X_3$$

$$C_{out} = X_1X_2 + X_2X_3 + X_1X_3$$

$$Carry = WX_4 + WC_{in} + X_4C_{in}$$

$$= WX_4 + WC_{in}(X_4 + X_4') + X_4C_{in}(W + W')$$

$$= WX_4 + WC_{in}X_4 + WC_{in}X_4' + X_4C_{in}W + X_4C_{in}W'$$

$$= WX_4(1 + C_{in} + C_{in}) + X_4'C_{in}W + X_4C_{in}W'$$

$$= WX_4 + C_{in}(X_4'W + X_4W')$$

$$WX_4 + C_{in}(W \oplus X_4)$$

If we consider WX_4 only:

$$WX_4 = W'X_4X_4' + WX_4X_4$$

$$= X_4(W'X_4' + WX_4)$$

$$= X_4(W \oplus X_4)'$$

Therefore, Carry is given by:

$$Carry = X_4(W \oplus X_4)' + C_{in}(W \oplus X_4)$$

$$Carry = X_4(X_1 \oplus X_2 \oplus X_3 \oplus X_4)' + C_{in}(X_1 \oplus X_2 \oplus X_3 \oplus X_4)$$

A.3 Expression for C_{out}

$$C_{out} = X_1X_2 + X_2X_3 + X_1X_3$$

$$= X_1X_2 + X_2X_3(X_1 + X_1') + X_1X_3(X_2 + X_2')$$

$$X_1X_2 + X_1X_2X_3 + X_1'X_2X_3 + X_1X_2X_3 + X_1X_2'X_3$$

$$X_1X_2\left(1 + X_3 + X_3\right) + X_3\left(X_1'X_2 + X_1X_2'\right)$$

$$= X_1X_2 + X_3(X_1 \oplus X_2)$$

$$X_1X_2 = X_1X_1'X_2' + X_1X_1X_2$$

$$= X_1(X_1'X_2' + X_1X_2)$$

$$= X_1(X_1 \oplus X_2)'$$

$$C_{out} = X_1(X_1 \oplus X_2)' + X_3(X_1 \oplus X_2)$$

Bibliography

[1] Parhami B 2010 *Computer Arithmetic: Algorithms and Hardware Designs* (New York: OUP)

[2] Preethy A P *et al* 2000 Low power CMOS pass logic 4-2 compressor for high-speed multiplication *43rd IEEE Midwest Symposium on Circuits and Systems* 3 pp 1296–8

[3] Parhi K K *et al* 2001 Low-power 4-2 and 5-2 compressors *Thirty-Fifth Asilomar Conference on Signals, Systems and Computers* vol 1 pp 129–33

[4] Veeramachaneni S, Krishna K, Avinash L, Puppala S and Srinivas M B 2007 Novel architectures for high-speed and low-power 3–2, 4–2 and 5–2 compressors, *20th Int. Conf. on VLSI Design held jointly with 6th Int. Conf. on Embedded Systems (VLSID'07)* vol 1 pp 324–9

[5] Chang C-H, Gu J and Zhang M 2004 Ultra low-voltage low-power CMOS 4-2 and 5-2 compressors for fast arithmetic circuits *IEEE Trans. Circuits Syst. I Regul. Pap.* **51** 1985–97

[6] Saha P, Samanta P and Kumar D 2016 4:2 and 5:2 Decimal Compressors *Proc. Int. Conf. Intell. Syst. Model. Simulation, ISMS* pp 424–9

[7] Han J 2016 Introduction to approximate computing *Proc. IEEE VLSI Test Symp.* 2016

[8] Choudhury M 2011 Approximate logic circuits: theory and applications Rice Univ.

[9] Jiang H, Liu C, Liu L, Lombardi F and Han J 2017 A review, classification, and comparative evaluation of approximate arithmetic circuits *ACM J. Emerg. Technol. Comput. Syst.* **13** 60

[10] Mittal S 2016 A survey of techniques for approximate computing *ACM Comput. Surv.* **48** 62

[11] Lingamneni A, Enz C, Nagel J L, Palem K and Piguet C 2011 Energy parsimonious circuit design through probabilistic pruning *Proc. Design, Autom. Test Eur. DATE* pp 764–9

[12] Lingamneni A, Enz C, Palem K and Piguet C 2011 Parsimonious Circuits for Error-Tolerant Applications through Probabilistic Logic Minimization *Integrated Circuit and System Design. Power and Timing Modeling, Optimization, and Simulation (Lecture Notes in Computer Science* vol 6951) (Berlin, Heidelburg: Springer) pp 204–13

[13] Mahdiani H R, Ahmadi A, Fakhraie S M and Lucas C 2010 Bio-inspired imprecise computational blocks for efficient VLSI implementation of soft-computing applications *IEEE Trans. Circuits Syst. I Regul. Pap.* **57** 850–62

[14] Allen C I, Langley D and Lyke J C 2015 Inexact computing with approximate adder application *Natl. Aerosp. Electron. Conf. Proc. IEEE* 2015 pp 21–8

[15] Venkatachalam S and Ko S B 2017 Design of power and area efficient approximate multipliers *IEEE Trans. Very Large Scale Integr. Syst.* **25** 1782–6

[16] Jiang H *et al* 2015 A Comparative Review and Evaluation of Approximate Adders *GLSVLSI '15: Great Lakes Symposium on VLSI* (ACM) pp 343–8

[17] Liang J, Han J and Lombardi F 2013 New metrics for the reliability of approximate and probabilistic adders *IEEE Trans. Comput.* **62** 1760–71

[18] Maddisetti L, Senapati R K and Ravindra JVR 2021 Accuracy evaluation of a trained neural network by energy efficient approximate 4:2 compressor *Comput. Electr. Eng.* **92** 107137

[19] Lavanya M and Ravindra J V R 2018 Performance metrics of imprecise multipliers based on proximate compressors for IIR filters *Proc. Int. Conf. Microelectron. ICM* 2018 pp 96–9

[20] Maddisetti L and Ravindra J V R 2019 Machine learning based power efficient approximate 4:2 compressors for imprecise multipliers *Proc. 32nd Int. Conf. on VLSI Design, VLSID 2019 - Held concurrently with 18th Int. Conf. on Embedded Systems, ES 2019* pp 221–6

[21] Yang Z, Yang J, Xing K and Yang G 2017 Approximate compressor based multiplier design methodology for error-resilient digital signal processing *2016 IEEE Int. Conf. Signal Image Process. ICSIP 2016* pp 740–4
[22] Samie F, Bauer L and Henkel J 2015 An approximate compressor for wearable biomedical healthcare monitoring systems *2015 Int. Conf. Hardware/Software Codesign Syst. Synth. CODES+ISSS 2015* pp 133–42
[23] Akbari O, Kamal M, Afzali-Kusha A and Pedram M 2017 Dual-quality 4:2 compressors for utilizing in dynamic accuracy configurable multipliers *IEEE Trans. Very Large Scale Integr. Syst.* **25** 1352–61
[24] Ha M and Lee S 2018 Multipliers with approximate 4-2 compressors and error recovery modules *IEEE Embed. Syst. Lett.* **10** 6–9
[25] Yi X, Pei H, Zhang Z, Zhou H and He Y 2019 Design of an energy-efficient approximate compressor for error-resilient multiplications *Proc. IEEE Int. Symp. Circuits Syst.* 2019-May 1–5
[26] Momeni A, Han J, Montuschi P and Lombardi F 2015 Design and analysis of approximate compressors for multiplication *IEEE Trans. Comput.* **64** 984–94
[27] Maheshwari N, Yang Z, Han J and Lombardi F 2015 A design approach for compressor based approximate multipliers *Proc. IEEE Int. Conf. VLSI Des.* 2015 pp 209–14
[28] Gorantla A and Deepa P 2017 Design of approximate compressors for multiplication *ACM J. Emerg. Technol. Comput. Syst.* **13** 44
[29] Chang Y J, Cheng Y C, Lin Y F, Liao S C, Lai C H and Wu T C 2019 Imprecise 4-2 compressor design used in image processing applications *IET Circuits, Devices Syst.* **13** 848–56
[30] Edavoor P J, Raveendran S and Rahulkar A D 2020 Approximate multiplier design using novel dual-stage 4:2 compressors *IEEE Access* **8** 48337–51
[31] Manikantta Reddy K, Vasantha M H, Nithin Kumar Y B and Dwivedi D 2019 Design and analysis of multiplier using approximate 4-2 compressor *AEU Int. J. Electron. Commun.* **107** 89–97
[32] Strollo A G M, Napoli E, De Caro D, Petra N and Meo G D I 2020 Comparison and extension of approximate 4-2 compressors for low-power approximate multipliers *IEEE Trans. Circuits Syst. I Regul. Pap.* **67** 3021–34
[33] Maddisetti L, Senapati R K and Ravindra J V R 2021 A neural network trained to perform as an imprecise 4:2 compressor using machine learning algorithm *IOP Conf. Series: Materials Science and Engineering* **1042** 12014
[34] Majumdar R and Saini S 2015 A novel design of reversible 2:4 decoder *2015 Int. Conf. on Signal Processing and Communication (ICSC)* pp 324–7

IOP Publishing

Advances in Image and Data Processing using VLSI Design, Volume 1
Smart vision systems
Sandeep Saini, Kusum Lata, Abhishek Sharma and G R Sinha

Chapter 14

A comparison of different procedures for hardware-based video shot boundary detection

Shrikant Chavate and Ravi Mishra

14.1 Introduction

Video shot boundary detection (SBD) is an essential step in tasks such as video indexing, browsing, and retrieval. The process of SBD can be achieved by segmenting a video into basic units known as shots. This chapter describes the different SBD techniques used to identify and differentiate the type of shot boundary, i.e. gradual or abrupt. It details the various methods which have been implemented to achieve SBD, along with their comparative analysis. Many advanced methodologies that use machine learning (ML) approaches and other traditional methods are framed here. The results, which were validated using Text REtrieval Conference Video Retrieval Evaluation (TRECVID) datasets, are also compared for different approaches. In addition, challenges such as motion, illumination, and lighting effects are discussed. A hardware implementation of SBD is covered in this chapter and various aspects of field-programmable gate array (FPGA)-based hardware systems are discussed.

In the field of engineering, video processing plays a vital role in large applications, which need to perform various manipulations of video streams (i.e. on frames or shots). High-definition video can require a lot of memory for storage, so compression techniques are necessary. Object tracking or object identification is also an area in which much considerable research has taken place and is still in progress.

The retrieval of visual content from big databases can also be achieved by video processing [1]. This can be a later stage of content-based image retrieval (CBIR). Video footage is composed of frames, which are combined to become a shot. The identification of shot(s) is a research area which has been found very useful for many real-time applications, such as content-based video retrieval (CBVR), in which the desired video content is retrieved from a database.

doi:10.1088/978-0-7503-3919-3ch14

We now discuss SBD in detail: by definition, a shot is a sequence of interconnected successive frames captured by a single camera which represents a continuous action in time and space. The automatic detection of the boundaries of a video shot is called SBD [2].

Figure 14.1 shows the structure of video; shot boundaries can be broadly classified into two major categories:

(i) those with abrupt (or hard) cuts, (ii) those with gradual (or soft) cuts. Gradual cuts or transitions can take different forms, such as fade in, fade out, dissolve, and wipe. An abrupt cut can be easier to detect than a gradual one, since gradual cuts consist of several special editing effects. If rapid change occurs between shots, the change is known as abrupt; in this case, considerable differences occur at the boundary between the shots. This effect is referred to as an abrupt transition. 'Fade in' is the effect of an image forming from darkness; such effects are usually used when a film begins. 'Fade out' is the opposite action to fade in; in this case, the image gradually changes into darkness. This can be seen at the ends of films. A dissolve is an effect in which one image starts to disappear, but at the same time, the next image starts to appear. Finally, a wipe is an effect in which pixels are replaced in an organized way, following, for example, circular, diagonal, vertical, or horizontal boundaries. The above mentioned effects [3] are shown in figure 14.2.

Video SBD tasks can face different issues, such as illumination effects, object movement, camera movement, etc.; these effects may lead to false detection.

The general procedure for SBD is based on the following few steps (described in [4]):

(i) Feature extraction
(ii) Formation of a continuity function
(iii) The detection and classification of the type of boundary

This process starts with feature extraction, as the features illustrate various aspects of the visual content of a video [5]. The feature variation from frame to

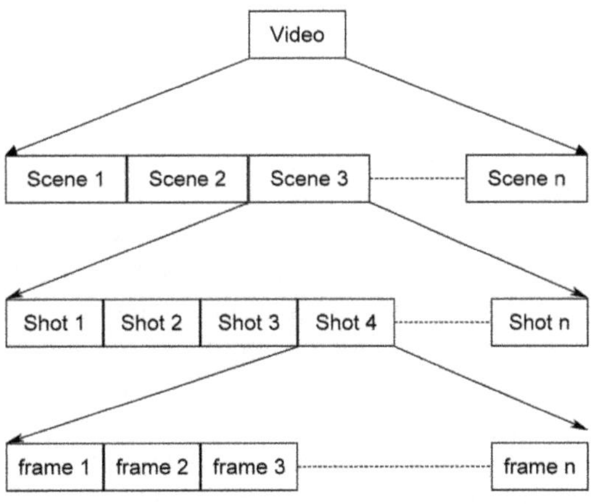

Figure 14.1. Basic video structure.

Figure 14.2. Examples of abrupt and gradual cuts. © 2019 IEEE. Reprinted, with permission, from [3].

frame then has to be analyzed. The difference in the discontinuity magnitude is used as the input of the detector. A threshold is then chosen, against which these values are compared. When the threshold is exceeded, a shot boundary is detected. Thus, the location of the shot boundary can be determined.

In the following section, we present a detailed literature review and a comparative analysis.

14.2 Literature review

Many techniques have developed over a number of decades and are still developing. Many of them have been reviewed and framed here, reflecting the state of the art. This includes the detection procedures, the capabilities of feature vectors, the effects of object motion and illumination, etc. At the same time, the contributions of advanced tools and techniques, such as machine learning, artificial intelligence, etc. are also framed.

SBD is a broad field; although a lot of work has been done so far, it is still an advanced research topic. Its objectives and implementation techniques are described here, along with a summary of its achievements (figure 14.3).

As previously mentioned, feature extraction, which is needed for video analysis and operations, acts as the input for the later stages of the procedure. Segmented video can be a platform for this, using which, frame-wise feature extraction is performed. A computation of the difference between frames has to be performed and the results used an the input of the detector, which are then compared with the threshold value. This comparison is used to determine the presence or absence of a shot boundary.

In this process, it is beneficial to make use of shot-length distribution, the visual discontinuity patterns at shot boundaries, and characteristic temporal changes of the visual features around a boundary. These techniques have advantages, such as robust and sequence-independent performance, as well as the simultaneous detection of different types of shot boundary [6]. Also, the different steps mentioned here that are used to carry out the above mentioned task of SBD include: detector development, hypothesis, false detection, missed detection, and the likelihood function [6]. One SBD technique is based on the just-noticeable difference (JND) color histogram, which is

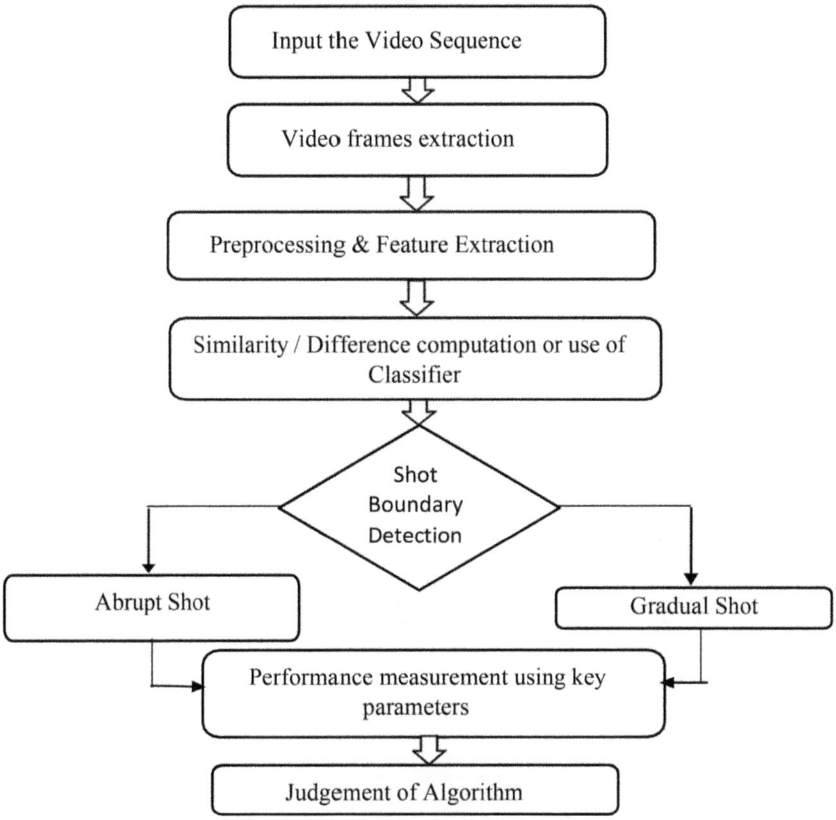

Figure 14.3. General flowchart of SBD.

called the just-noticeable histogram. In this method, two consecutive frames have to be chosen, and the histogram is computed. Then, for a pair of histograms, a measure of similarity can be calculated. This can be used as an indicator, and is based on an adaptive-threshold sliding window [2]. In the JND approach, colors are defined based on vector similarity. The degree of similarity is given by:

$$\mathrm{Simi}_{\mathrm{JND}(x,y)} = \sum_{i=1}^{c} \min\left(h_x(i),\, h_y(i)\right), \tag{1}$$

where h_x, h_y are the JND histograms for frames x and y, and c is the number of color bins in the histogram. $\mathrm{Simi}_{\mathrm{JND}(x,y)}$ is the total pixel count present in both frames being compared.

Yan [7] *et al* presented an approach based on speeded up robust features (SURF) and kernel canonical correlation analysis (KCCA) for the robust registration of remote sensing images. It claimed to be able to detect false matches.

SURF encodes the distribution of pixel intensities in the neighborhood of the detected features, rather than the row pixel intensities [7]. This proposed approach focused on the following few steps:

i) Initial feature matching using SURF
ii) Criterion for false match rejection
iii) Influence function
iv) False match rejection procedure

Zheng [8] *et al* explained the use of a combination of features and a support vector machine (SVM) to detect abrupt cuts. In this paper, the issues of the presence of large objects, camera movements, and flash lights were discussed; here, global features were claimed to be insufficient to detect abrupt transition (AT) boundaries. This paper recommended a combination of local and global features for better performance. Also, real-time SBD using unsupervised clustering techniques [9] was claimed to achieve improved accuracy using LAB features and edge features.

Hannane *et al* [10], presented a work on SBD using the scale invariant feature transform-point distribution histogram (SIFT-PDH) approach, in which features were extracted by combining both types of feature. The difference between each pair of consecutive frames of the video was found and then a difference comparison was obtained by comparing SIFT-PDHs. This proposed system is shown in figure 14.4. Fan *et al* [11] described fuzzy color distribution chart-based SBD, which was used to describe the spatial distribution of colors and reduce the issues of poor illumination and the insertion of logos (figure 14.5). It was also claimed that this method can

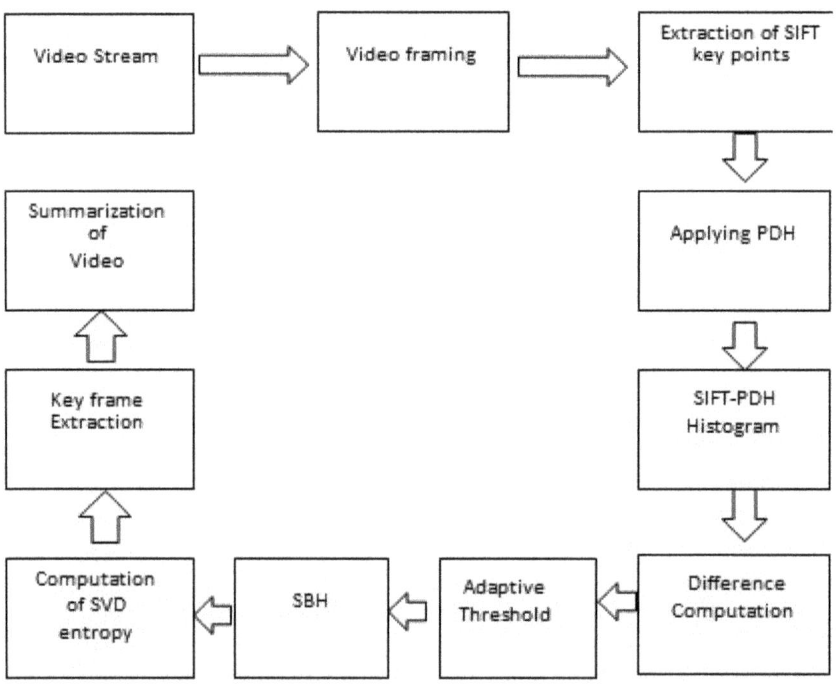

Figure 14.4. Approach used for SIFT-PDH. Reprinted with permission from [10]. Copyright 2016 Springer.

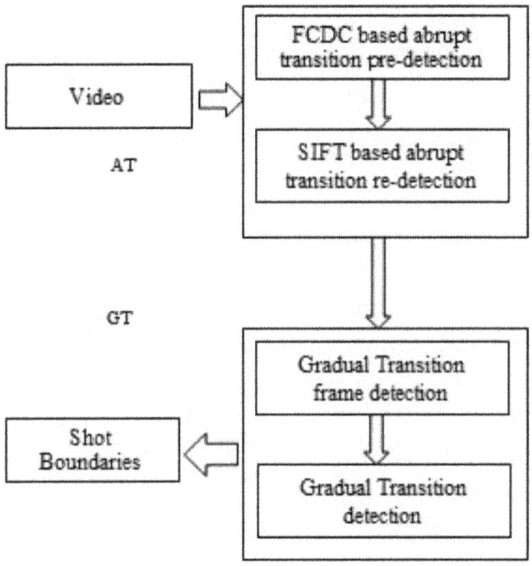

Figure 14.5. The fuzzy color distribution chart approach. Reprinted with permission from [11]. Copyright 2017 Springer.

distinguish gradual cuts when object movements are present. This also helps to address some SBD issues that remained have unsolved elsewhere.

Furthermore, Mishra *et al* [5, 12] proposed and implemented real- and non-real-time SBD using dual tree complex wavelet transform (DTCWT). For non-real-time videos, they claimed that the use of DTCWT and nodal analysis concepts led to efficient results and better accuracy. Methods such as the calculation of differences between frames, standard deviation, and histogram differences were used to detect abrupt as well as gradual cuts. Detection was also performed in the presence of illumination and object movement [13], and this approach was compared to the block matching algorithm (BMA) and CWT algorithms.

Wu *et al* [14] presented research into SBD using SURF features. Here, only the half of the frames were considered for SURF feature extraction, which reduced the execution time. This was followed by a feature matching algorithm and computation of a similarity measure. Wu *et al* [3] worked on SBD using a two-stage detection procedure; separate detections were made for abrupt and gradual transitions. The abrupt-cut detection made use of a fusing color histogram and deep features, while for gradual cuts, 3D convolutional neural network (C3D)-based analysis was used. They claimed to make efficient use of a spatiotemporal convolutional neural network (CNN). In [15] the use of a fully convolutional CNN architecture was also claimed to be faster in the detection of shot boundaries.

Opera *et al* [16] proposed and implemented an SBD method for low-complexity high efficiency video coding (HEVC) encoders, in which chroma-subsampled components were used for faster results. Gao *et al* [4] worked on SBD using visual perception; here, redundant frames first had to be removed, then the visual consistency features of the video were used for further SBD tasks.

Shen *et al* [17] proposed and implemented a hybrid method for boundary detection using a combination of High Level Fuzzy PetriNets (HLFPN) and keypoint matching. In addition, histogram difference and SURF features were used. They claimed that this algorithm worked effectively with illumination changes.

Bi *et al* [18] proposed and implemented a dynamic mode-decomposition approach for SBD, which extracted some temporal foreground modes and one temporal background mode from video; detection took place at sharp changes of amplitude.

Karpagavalli *et al* [19] combined a Hessian matrix with the minimum eigenvalues of the point of interest and the region of interest, respectively. They carried out a comparison of the performance of their method and those of earlier methods.

Yuan *et al* [20] presented a formal study of SBD and a graph partitioning model and performed a comparison of active and random SVMs.

Park *et al* [21] explained the use of an adaptive threshold to benefit from improved precision and recall values.

Lakshmi Priya *et al* [22] used feature vectors such as color, edge, texture, and motion association in conjunction with a Walsh–Hadamard transform (WHT) kernel and a WHT matrix. Here, a combination of weighted feature vectors was used for the formation of a single continuity value, which was then used for further detection procedures.

Kim *et al* [23] proposed a technique for gradual cut detection based on motion and illumination effects; an adaptive threshold helped to identify multiframe motion-based gradual cuts. Some approaches, for example, statistical methods [24] have used features such as the mean and the standard deviation to calculate frame differences. These were found to be easy to compute and simple.

Kar and Kanungo [25] described an SBD based on the local binary pattern (LBP) feature. The texture features had to be extracted from LBP and it was claimed that this method performed better by reducing false detection.

Yazdi and Fani [26] worked on AT and GT detection using a frame histogram and an adaptive threshold. For GTs, a *k*-means classifier was used, followed by singular value decomposition (SVD) and SVM.

Zheng and Zhang [27] suggested an algorithm that used a hue, saturation, value (HSV) histogram and SURF, which were accelerated using a graphics processing unit (GPU). The SVM helped in the detection of abrupt cuts.

14.3 Methodology

Following our review of various literature, we decided that the proposed system must include an implementation of basic steps, such as the preprocessing of video frames in a stream of video. This should be followed by the application of precise matching techniques to the frames by reference to a comparison of several techniques, and thus a choice of the most efficient one to achieve the preliminary tasks. The next step is the selection of either a type of threshold or a classifier with which to find matched frames. Frames from video databases are analyzed using most the favorable tool(s).

14.3.1 Comparative discussion of results

The TRECVID dataset is preferred for work on video (SBD) tasks. Video shot boundary detection (VSBD) can be implemented effectively using MATLAB, OpenCV, or Python. The built-in MATLAB library has helped many experiments to be performed effectively using the TRECVID dataset. Various tools and software are also available to carry out VSBD tasks.

The comprehensive evaluation criteria used to validate and estimate the performance of VSBD are the precision (P), the recall rate (R), and the combined score (F1 score). They are given by the following formulae:

$$\text{Precision (P)} = \frac{\text{Correct transitions}}{\text{Total transitions}} * 100$$

$$\text{Recall (R)} = \frac{\text{Correct transitions}}{\text{Actual transitions}} * 100$$

$$\text{F1 score} = 2 * \frac{\text{P} * \text{R}}{(\text{P} + \text{R})} * 100.$$

Table 14.1 shows a comparison of different methods and their results with respect to the dataset(s) adopted for experimentation (see figure 14.6).

14.3.2 Hardware implementation

A hardware implementation of SBD is an effective way to recognize the occurrence of shot boundaries in real-time applications. An FPGA-based hardware detection platform was proposed and implemented in [28]; this used the local histogram method followed by a computation of the dissimilarity between consecutive frames.

Table 14.1. Comparison of results for various approaches.

Label	Method (M)	Precision (%)	Recall (%)	F1 (%)
M1	Unsupervised clustering	98.04	98.04	98.04
M2	WHT	89.30	86.30	87.70
M3	Global and local feature descriptors	88.50	65.90	74.60
M4	Supervised clustering	94	69	82
M5	Fast algorithm for SBD with SURF	100	99.84	99.91
M6	HLFPN and keypoint matching	86.56	72.39	76.90
M7	Time warping and mean shift	95.40	100	97.50
M8	Blocked HSV	95.70	86.10	96.60
M9	C3D	81.30	79.60	80.40
M10	HSV+DPHA	90.70	90.10	90.40
M11	TSSBD	93.20	93.80	93.50
M12	GPU+SURF	90.35	63.59	72.60
M13	Graph model	96.47	94.41	95.37
M14	Combined features and SVM	92.84	75.87	82.10

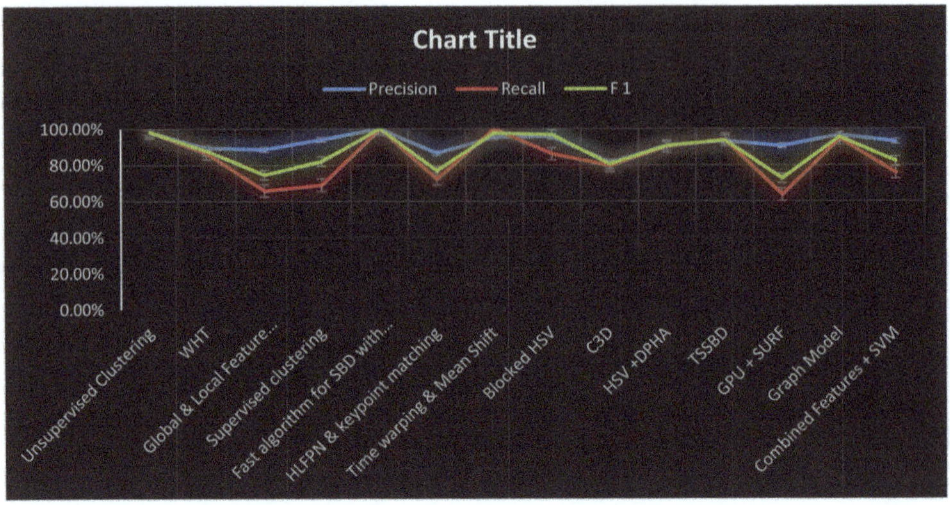

Figure 14.6. Graphical representation of precision, recall, and F1.

The advantage of using FPGAs is that they are a compromise between ASICs and general-purpose processors; it is also claimed that they offer high performance. The authors of [28] used the following hardware configuration: an XSV800 board containing a Virtex XCV800 FPGA core with 800 000 gates, 9408 slices, 18816 flip flops, 18816 LUTs (lookup tables), 320 IOBs (input/output blocks), 4 GClks (generic clocks) and 28 configurable internal block random-access memories (RAMs), each of 4 KB. The architecture is described using the VHSIC hardware description language (VHDL) and register-level transfer (RTL) in the Integrated Software Environment (ISE) of Xilinx. Figures 14.7 and 14.8 show schematics for cut detection and entire image video, respectively. In figure 14.7, adjacent frames are considered, which are shown divided into 16 blocks. These are used for cut detection with the selected threshold. Figure 14.8 shows a schematic of a video in which the visible and invisible pixels are shown separately with their respective counts. Figure 14.9 shows the design for cut detection at the RTL level. Table 14.2 presents the results obtained using the FPGA-based method implemented in [28]. This was claimed to achieve a minimum period of 8.488 ns.

Another FPGA-based hardware SVM implementation for SBD was described in [29]; this paper claimed that the throughput was increased by the use of a pipelined architecture in the feature extraction stage. It also claimed that hardware SVM was able to detect abrupt and gradual transitions in the video stream. Random pseudo-sampling techniques were employed that helped to solve the class imbalance problem in SVM training. The threshold method in the postprocessing stage merged small subshots, which reduced false alarms. The complete system, built on a Xilinx Virtex IV XC4VSX35 FPGA platform, achieved 256 frames per second.

Figure 14.10 shows the hardware SVM architecture; a linear kernel was used here, which offered the advantage of minimal hardware complexity. In this

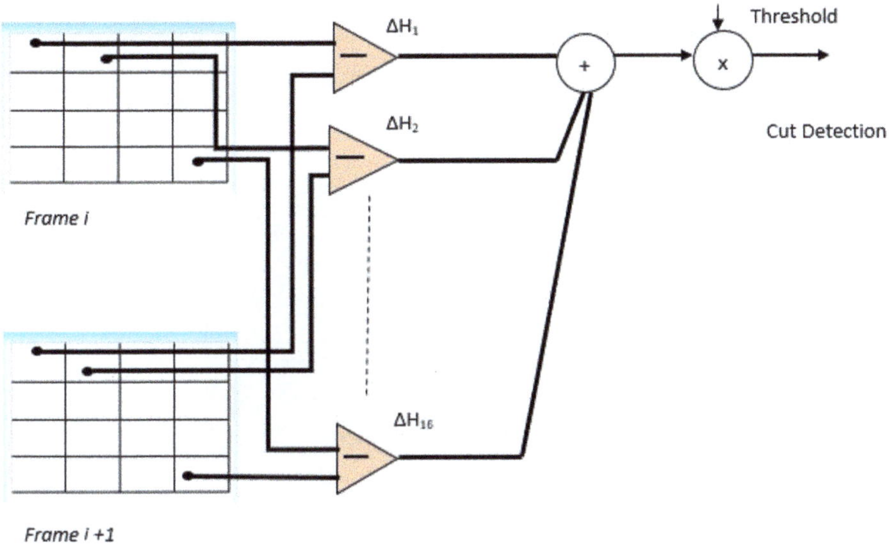

Figure 14.7. Cut detection. Reprinted from [28]. Copyright 2007, with permission from Elsevier.

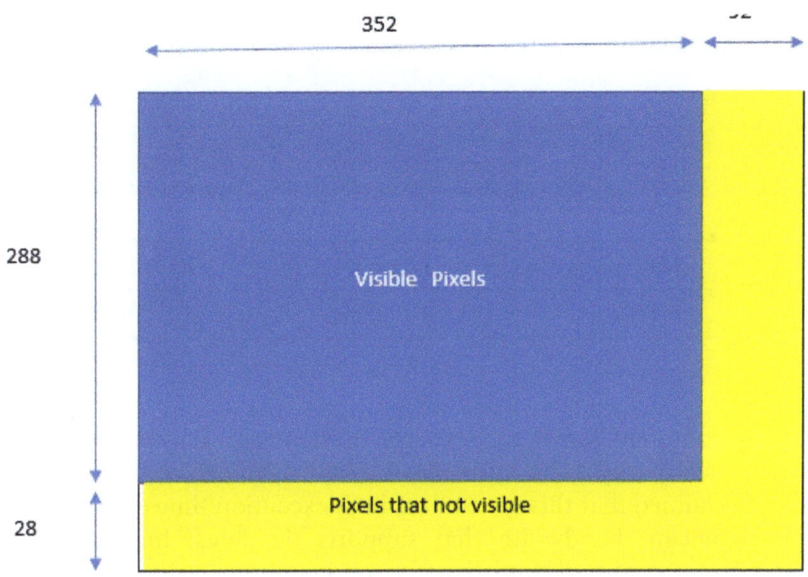

Figure 14.8. Entire image video. Reprinted from [28]. Copyright 2007, with permission from Elsevier.

architecture, false alarms were reduced by merging the time distances of shots that were lower than a threshold.

Another local histogram (LH)-based approach that included an FPGA-based SOC in a hardware implementation of SBD is described in [30]. The modules

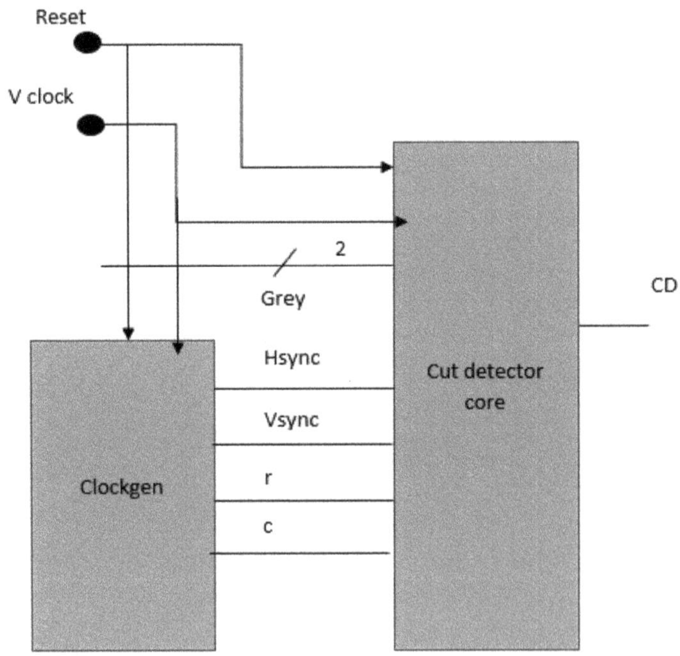

Figure 14.9. Synoptic of the design at the RTL level. Reprinted from [28]. Copyright 2007, with permission from Elsevier.

Table 14.2. Results of the FPGA-based approach. Reprinted from [28]. Copyright 2007, with permission from Elsevier.

Slices	Min. period	Flip flops	IOBs	LUTs
152 (1%)	8.488 ns	112 (0.6%)	4 (1%)	268 (1%)

developed had different hardware resource occupations; it was claimed that they were used in such a way as to allow flexible management of the target hardware system. It was claimed that this system had high execution time efficiency and that it reached a significant bandwidth that supports the latest high-resolution video formats at a large number of frames per second.

Figure 14.11 shows a method for the detection of abrupt cuts; the LH was tested on different grayscale quantization levels. The authors of [30] claimed that this system provided efficient performance for HD video formats in real time. Table 14.3 shows the hardware resource occupation [31] and table 14.4 gives the frequency results for a counter-based architecture and a memory-based architecture [31], respectively.

A study of color structure descriptors for temporal video segmentation proposed in [32] paid attention to real-time analysis and presented a hardware design for faster

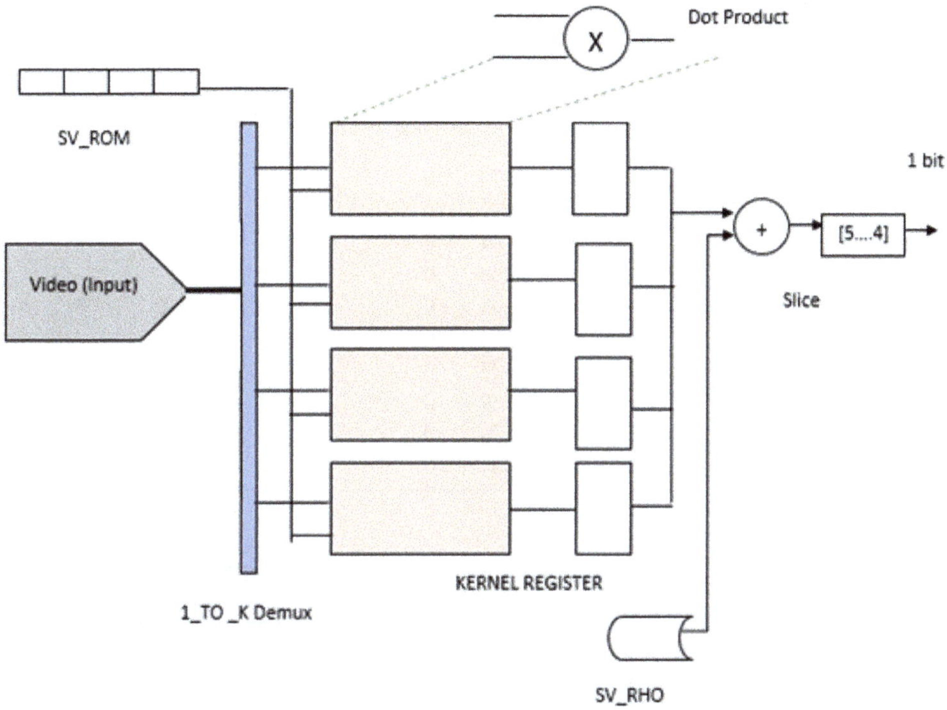

Figure 14.10. Hardware SVM architecture. © 2019 IEEE. Reprinted, with permission, from [29].

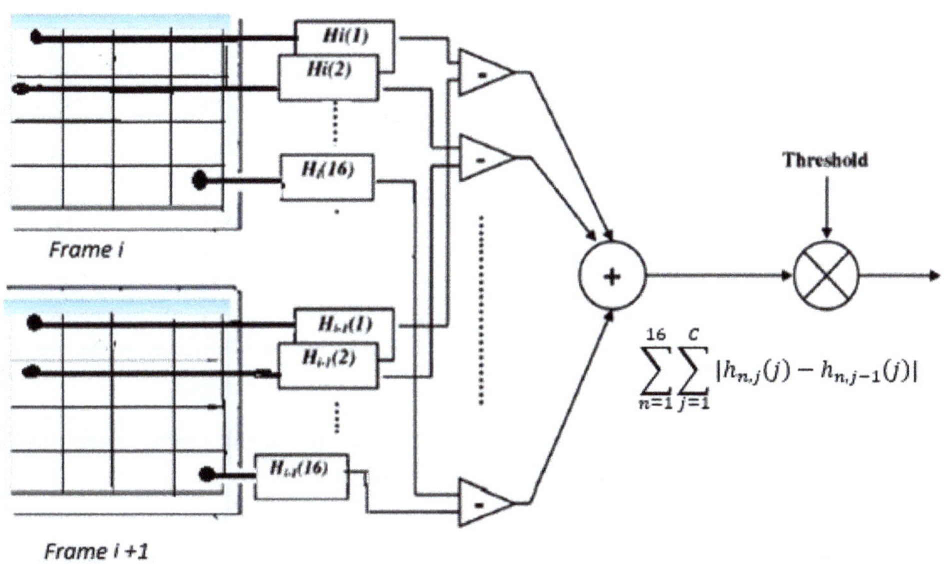

Figure 14.11. Abrupt transition detection using the LH technique. Reproduced from [30] with permission of TÜBİTAK.

Table 14.3. Hardware resource occupation for a register-based architecture. Reproduced from [30] with permission of TÜBİTAK.

Image size	Quantization levels (C)	Slice registers	Slice LUTs	BRAM 18K	BRAM 36K	DSP48Es	Frequency
320 × 240	4	957	1079	0	0	1	341
	8	1790	1979	0	0	1	341
512 × 512	4	1022	1144	0	0	1	318
	8	1919	2108	0	0	1	318
640 × 480	4	1086	1208	0	0	1	317
	8	2047	2236	0	0	1	317
1280 × 1024	4	1217	1333	0	0	1	307
	8	2306	2489	0	0	1	307
1920 × 1080	4	1217	1328	0	0	1	305
	8	2306	2484	0	0	1	305

Table 14.4. Hardware resource occupation for a memory-based architecture. Reproduced from [30] with permission of TÜBİTAK.

Image size	Quantization levels (C)	Slice registers	Slice LUTs	BRAM 18K	BRAM 36K	DSP48Es	Frequency
320 × 240	4	123	191	1	0	1	341
	8	124	192	1	0	1	341
	256	129	194	1	1	1	224
512 × 512	4	124	192	1	0	1	318
	8	125	193	1	0	1	318
	256	130	195	0	2	1	222
640 × 480	4	124	193	1	0	1	317
	8	125	194	1	0	1	317
	256	130	196	0	2	1	221
1280 × 1024	4	127	191	1	0	1	307
	8	128	192	1	0	1	307
	256	133	193	0	2	1	219
1920 × 1080	4	127	188	1	0	1	305
	8	128	189	1	0	1	305
	256	133	177	0	2	1	219

execution. It was claimed that this design permitted the simultaneous execution of multiple treatments by providing suitable memory sharing.

14.4 Conclusions

As per the various methodologies available for VSBD, the machine learning, CNN, and AI-based approaches offer flexible ways to achieve accurate results. False alarms can also be controlled using deep learning methods. FPGA-based hardware implementations offer significant performance for SBD.

Acknowledgements

We acknowledge the researchers who have contributed to the field of video processing, in particular, those who have worked on video shot boundary detection.

Bibliography

[1] Sengupta A, Thounaojam D M, Singh K M and Roy S 2015 Video shot boundary detection: a review *Proc. 2015 IEEE Int. Conf. Electr. Comput. Commun. Technol. ICECCT 2015*

[2] Janwe N J and Bhoyar K K 2013 Video shot boundary detection based on JND color histogram *2013 IEEE 2nd Int. Conf. Image Inf. Process. IEEE ICIIP 2013* pp 476–80

[3] Wu L, Zhang S, Jian M, Lu Z and Wang D 2019 Two stage shot boundary detection via feature fusion and spatial-temporal convolutional neural networks *IEEE Access* **7** 77268–76

[4] Gao Y, Lai Y and Liu Y 2019 Fast video shot boundary detection based on visual perception *2019 IEEE Int. Conf. Consum. Electron. ICCE 2019* pp 1–4

[5] Mishra R, Raman C V, Singhai S K and Sharma M 2015 Real time and non real time video shot boundary detection using dual tree complex wavelet transform *2015 Int. Conf. Ind. Instrum. Control. ICIC 2015* pp 1495–500

[6] Hanjalic A 2002 Shot-boundary detection: unraveled and resolved? *IEEE Trans. Circuits Syst. Video Technol.* **12** 90–105

[7] Yan W, She H and Yuan Z 2014 Robust registration of remote sensing image based on SURF and KCCA *J. Indian Soc. Remote Sens.* **42** 291–9

[8] Youxian Z and Yuan Z 2017 Abrupt shot boundary detection with combined features and SVM *Proc. 2016 2nd IEEE Int. Conf. Comput. Commun. ICCC 2016* pp 409–13

[9] Su N, Zhang J, Zhang Y and Zhang G 2019 Unsupervised clustering based real-time shot boundary detection for live broadcasting *2019 IEEE 5th Int. Conf. Comput. Commun. ICCC 2019* pp 135–40

[10] Hannane R, Elboushaki A, Afdel K, Naghabhushan P and Javed M 2016 An efficient method for video shot boundary detection and keyframe extraction using SIFT-point distribution histogram *Int. J. Multimed. Inf. Retr.* **5** 89–104

[11] Fan J, Zhou S and Siddique M A 2017 Fuzzy color distribution chart-based shot boundary detection *Multimed. Tools Appl.* **76** 10169–90

[12] Mishra R, Singhai S K and Sharma M 2014 Comparative study of block matching algorithm and dual tree complex wavelet transform for shot detection in videos *Proc. Int. Conf. Electron. Syst. Signal Process. Comput. Technol. ICESC 2014* pp 450–5

[13] Mishra R, Singhai S K and Sharma M 2013 Video shot boundary detection using dual-tree complex wavelet transform *Proc. 2013 3rd IEEE Int. Adv. Comput. Conf. IACC 2013* pp 1201–6

[14] Wu Z and Xu P 2014 A fast gradual shot boundary detection method based on SURF *Practical Applications of Intelligent Systems (Advances in Intelligent Systems and Computing* vol 279*)* ed Z Wen and T Li (Berlin: Springer)

[15] Gygli M 2018 Ridiculously fast shot boundary detection with fully convolutional neural networks *2018 Int. Conf. on Content-Based Multimedia Indexing (CBMI) (La Rochelle)* pp 1–4

[16] Oprea C C, Preda R O, Pirnog I and Al Dobre R 2019 Video shot boundary detection for low complexity HEVC encoders *Proc. 10th Int. Conf. Electron. Comput. Artif. Intell. ECAI 2018* pp 1–4

[17] Shen R, Lin Y, Juang T T, Shen V R L and Lim S Y 2018 Automatic detection of video shot boundary in social media using a hybrid approach of HLFPN and keypoint matching *IEEE Trans. Comput. Soc. Syst.* **5** 210–19

[18] Bi C *et al* 2018 Dynamic mode decomposition based video shot detection *IEEE Access* **6** 21397–407

[19] Karpagavalli S, Balamurugan V and Kumar S R 2020 A novel hybrid keypoint detection algorithm for gradual shot boundary detection *Int. Conf. Emerg. Trends Inf. Technol. Eng. IC-ETITE 2020* pp 1–5

[20] Yuan J *et al* 2007 A formal study of shot boundary detection *IEEE Trans. Circuits Syst. Video Technol.* **17** 168–86

[21] Park S, Son J and Kim S J 2017 Effect of adaptive thresholding on shot boundary detection performance *2016 IEEE Int. Conf. Consum. Electron. ICCE-Asia 2016* pp 1–2

[22] Lakshmi Priya G G and Domnic S 2014 Walsh-hadamard transform kernel-based feature vector for shot boundary detection *IEEE Trans. Image Process.* **23** 5187–97

[23] Kim S, Hong H and Nang J 2015 A gradual shot change detection using combination of luminance and motion features for frame rate up conversion *2015 11th Int. Conf. on Signal-Image Technology & Internet-Based Systems (SITIS) (Bangkok)* pp 295–9

[24] Madhusudhan M V and Hegde C 2015 Video shot boundary detection using statistical methods *2015 IEEE Int. WIE Conf. on Electrical and Computer Engineering (WIECON-ECE) (Dhaka)* pp 52–6

[25] Kar T and Kanungo P 2015 A texture based method for scene change detection *2015 IEEE Power, Communication and Information Technology Conf. (PCITC) (Bhubaneswar)* pp 72–7

[26] Yazdi M and Fani M 2016 Shot boundary detection with effective prediction of transitions' positions and spans by use of classifiers and adaptive thresholds *2016 24th Iranian Conf. on Electrical Engineering (ICEE) (Shiraz)* pp 167–72

[27] Zheng Y and Zhang Y 2016 GPU-accelerated abrupt shot boundary detection *2016 16th Int. Symp. on Communications and Information Technologies (ISCIT) (Qingdao)* 141–5

[28] Boussaid L, Mtibaa A, Abid M and Paindavoine M 2007 A real-time shot cut detector: hardware implementation *Comput. Stand. Interfaces* **29** 335–42

[29] Hsu C F, Ku M K and Liu L Y 2009 Support vector machine FPGA implementation for video shot boundary detection application *Proc. IEEE Int. SOC Conf. SOCC 2009* pp 239–42

[30] Ben Abdelali A, Krifa M N and Mtibaa A 2017 FPGA-based SOC for hardware implementation of a local histogram-based video shot detector *Turkish J. Electr. Eng. Comput. Sci.* **25** 3300–18

[31] Ben Abdelali A, Krifa M N, Touil L, Mtibaa A and Bourennane E 2009 A study of the color-structure descriptor for shot boundary detection *IJ-STA—Int. J. Sci. Tech. Autom. Control Comput. Eng.* **3** 956–71 (Available: https://sta-tn.com/IJ_STA/Papers/volume_3N1_jul_2009/P8_IJSTA_V3N1_09.pdf)

[32] Ben Abdelali A, Hannachi M, Touil L and Mtibaa A 2017 Adequation and hardware implementation of the color structure descriptor for real-time temporal video segmentation *J. Real-Time Image Process.* **13** 739–58

IOP Publishing

Advances in Image and Data Processing using VLSI Design, Volume 1
Smart vision systems
Sandeep Saini, Kusum Lata, Abhishek Sharma and G R Sinha

Chapter 15

Hardware–software co-simulation of vehicle license plate detection on the ZedBoard SoC platform

Surbhi Chhabra, Sandeep Saini and Kusum Lata

15.1 Introduction

Vehicle license plate detection systems (VLPDSs) have attracted much interest from image processing researchers. VLPDSs are real-time programs that can process license plates in various environments and with plate variations, including lighting, background, position, numbers, size, and other factors. With more cars on the road, manually managing traffic regulation for smooth traffic flow is becoming increasingly complex, and traffic management needs to be improved. As a result, a reliable and efficient vehicle identification system is needed. Each country's vehicles have an individual and unique license number shown on their license plate. This number distinguishes one car from another, which is very important; number plates determine the parking time in parking garages. The number plate is electronically detected and recorded in the database when a vehicle enters a gate, as shown in figure 15.1 [1].

Vehicle recognition is a hot topic in the development of intelligent automated traffic systems and surveillance. Due to ethnic disputes and the current security situation, this is an area in which the creation of gadgets that can be utilized in various conditions to alleviate security worries is critical. Visually inspection of each car's license plate is difficult and time-consuming for a police officer. Furthermore, employing multiple officers to work as full-time license plate inspectors is not feasible. Inspectors cannot just drive around in their automobiles, gazing at license plates [28]. A mechanism is needed that can detect and identify license plates without continual human interaction. To address this challenge, artificial intelligence systems have been developed to identify license plates and retrieve the characters that are

doi:10.1088/978-0-7503-3919-3ch15

Figure 15.1. Vehicle license plate detection. Reproduced from [29] © 2021 The Author(s). Published by IOP Publishing Ltd. CC BY 4.0.

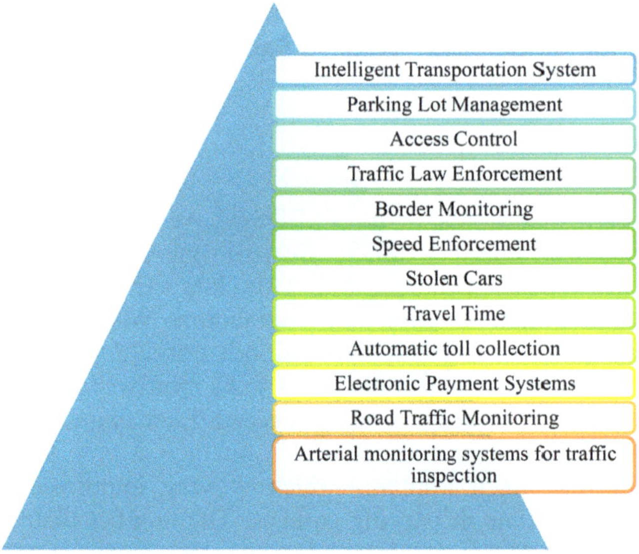

Figure 15.2. The wide range of applications of VLPDSs.

present on them. Further information about the car and its owner may be obtained using the license plate number, which can be utilized for further processing. As a result, there is a pressing need to create VLPDSs [2, 3]. Figure 15.2 depicts the wide variety of applications for VLPDSs.

Field-programmable gate arrays (FPGAs) are frequently utilized to build applications that need high-speed parallel data processing, such as image processing, and are widely employed in FPGA implementations. Image processing on a computer enables the manipulation of pixels inside a digital image for various reasons. The image processing approach, which is extensively utilized in various fields, such as medical, video surveillance, target identification, and robotics applications, improves the quality of a picture. Because real-time image processing is time-consuming, we must implement it at the hardware level, which provides parallelism and dramatically decreases the processing time [4, 5].

15.1.1 Literature survey

The retrieval of content from images and videos has sparked a lot of interest in recent computer vision and pattern recognition research. Objects, shape, texture, color, and the interactions between them are all examples of content. A novel and fast vertical edge detection method (VEDA) has been presented for license plate extraction [6, 7]. Content-based image retrieval, indexing, and classification can benefit from the semantic information supplied by a picture [6–8]. Textual data are intriguing, because they may be utilized to rapidly and convey the contents of an image, as mentioned by the authors of [6]. Textual data may be presented in an image or video in a variety of font styles, sizes, orientations, colors, and against a complex background, making extraction of the appropriate text section challenging. Furthermore, existing optical character recognition (OCR) methods can only recover text from a simple monochromatic backdrop and cannot extract text from a complex or textured background [9]. The algorithm developed in [10, 11] leverages the focus-of-attention mechanism from visual perception to recognize text areas.

Various approaches for detecting and localizing text in pictures and videos have been proposed in the past. These methods consider a variety of features linked to the text in a picture, including color, intensity, connected components, edges, and so on. These characteristics differentiate text regions from the image's backdrop and other locations. Color clustering is the basis of the method presented by the authors of [12]. If there is any noise in the input image, it is first preprocessed to eliminate it. The picture is then separated into color layers and a gray component. This method uses the fact that the color data in text characters differs from the color data in the backdrop. These layers' linked component-based algorithms are used to locate probable text areas. In addition, each row and column value is analyzed using the aligning and merging analysis (AMA) technique [12]. According to these studies, the algorithm is resilient when used to find primarily Chinese and English characters in pictures; nevertheless, some false alarms occurred due to uneven lighting or reflections in the test images. Color continuity is also used in the text detection method described in [13], which additionally employs multiresolution wavelet transformations for text region extraction and integrates low- and high-level picture characteristics. Except for minor text characters that are not accurately recognized, the findings demonstrate that the method is resilient in most instances. Misclassifications in texture segmentation also occur when an image has low contrast. The authors of [11] employed spatial variance and intensity profiles to recognize the text areas in pictures. A Gaussian pyramid was generated from the original images, which had various resolutions and sizes. The highest-resolution image and every lower-resolution image within the pyramid captured the text areas. Edge-based methods, morphological procedures, and picture projection profiles were first used to recognize text in multiscale images [10]. Wavelet features and the support vector machine (SVM) were used to verify the identified text areas; these algorithms are resistant to variations in font color, size, and language.

Text data are intriguing, because they may be used to describe the contents of a picture fast and clearly [14]. Recent research that employed extracted text has discovered a wide range of uses. One such newly created application is the mobile banking application supplied by financial institutions, which allows users to complete transactions, even if they only have the image of a check to send to the server [15]. Another use is a translation camera, which collects pictures, identifies text, and then translates it into the required language from the local language. If text-to-speech (TTS) software is utilized inside text detection and localization recognition, which is known as OCR, this can also be beneficial for visually impaired individuals [16]. In [17], the authors used a connected-component (CC)-based approach to identify and recognize the writing on cargo containers under varying illumination conditions and with characters of various sizes and forms. After quantizing an input picture, the difference between neighboring pixels was utilized to define the boundaries of possible letters. The pixels on the boundaries were then used to select local threshold values for each text candidate. These potential characters were utilized to produce CCs with the same gray level. Multiple algorithms based on aspect ratio, contrast histogram, and run-length measurements were then employed to filter out the non-text components. The authors of [18] created an algorithm that can be used to monitor and manage parking services in car parks. The authors of [20] employed a texture-based method and worked with edge information for localization and recognition. References [7, 15, 20, 21] presented a unique license plate extraction approach called histogram-based edge processing. In this method, vertical and horizontal histograms are produced, after which, they are sent through a low-pass filter to complete license plate segmentation. The methods utilized in [22, 23] are edge-based text extraction methodologies, whereas the approach used in [24] is based on connected components. The results were dependent on the environmental and lighting conditions, distance from the camera (scale), rotation, color, and the horizontal and vertical alignments of text in an image. Experiments were also conducted on images that included a variety of font styles and text characters from languages other than English.

FPGAs play an important role in hardware implementations. FPGAs are frequently utilized in image processing applications that require high-speed parallel processing. FPGAs have shown to be helpful in applications involving massive, independent data sets and digital data collection. Digital signal processing (DSP) is one of the commonest applications of FPGAs. Application-specific integrated circuits (ASICs) have traditionally been used to implement DSP. However, due to their lower cost and reconfigurability, FPGAs are increasingly replacing them. The potential uses of FPGAs have grown as chip densities have increased according to Moore's law. The most recent boards have several I/O ports, analog-to-digital (AD) and digital-to-analog (DA) converters, and a soft or hard core microprocessor, allowing the device to function as a system-on-chip (SoC). Furthermore, software/hardware co-design is gaining traction as an answer to some of the most complicated and time-consuming computations that FPGAs can perform. System Generator is a Xilinx DSP design tool that permits FPGA creation using the Mathworks Simulink model-based design environment. It is a system-level modeling tool that uses a Xilinx-specific block set to record DSP-friendly Simulink-modeled environment designs. System Generator has several features, including system resource estimation, which forecasts the use of FPGA resources, hardware co-simulation [25], and faster simulation through hardware-in-the-loop co-simulation, which can boost the simulation speed by many orders of magnitude [26]. Consequently, we aim to use the Xilinx System Generator (XSG), a program with a high-level graphical interface that runs under Matlab and leverages Simulink-based blocks to form various easy-to-handle applications for hardware description [5, 27].

15.1.2 Contributions and organization

The main contributions of this chapter are:

- We first describe three methodologies for the detection of license plates implemented in the MATLAB software and analyze the results.
- We also present the Simulink model and then the hardware–software co-simulation model for the VLPDS using XSG and analyze the experimental and simulation results.
- The results show that the VLPDS system utilizes minimum hardware resources and exhibits maximum throughput.

Organization of this chapter: the remaining sections of this chapter are arranged as follows: section 15.2 contains a thorough discussion of license plate extraction methods. In section 15.3, we present the simulation results. In section 15.4, we describe the proposed hardware–software co-simulation implementation. An overview of usage and the experimental findings can be found in section 15.5. Section 15.6 offers the conclusions and lists possible future projects.

15.2 Descriptions of algorithms

15.2.1 Edge-based extraction

The following are the basic phases of the edge-based extraction method, as shown in figure 15.3.

Detection: creation of a Gaussian pyramid, directional kernels, and feature map: in this technique, given an input image, the regions of the image that have the potential to contain text are recognized. The input picture is downsampled by half in each direction and filtered with a Gaussian kernel with a size of 3 × 3 to yield a Gaussian

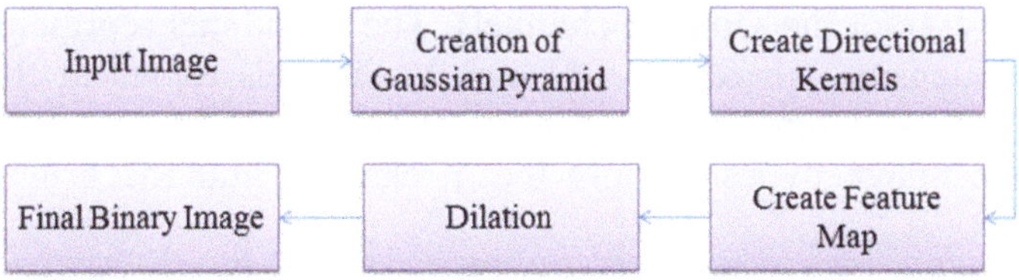

Figure 15.3. Flow chart (edge based).

pyramid. 'Downsampling' is the process by which a picture is scaled to a lower resolution than its original resolution. The resolution of a given image matches a level of the pyramid. These images are convolved with directional filters using kernels for different orientations to perform edge detection in the horizontal (0 degree), vertical (90 degree), and diagonal (45 degree, 135 degree) directions. The image is convolved with the orientation kernels to make a feature map. In the next step, a weighting factor is attached to each pixel to describe whether it is a contender or non-candidate for the text region. The pixel is considered to be a contender if it is highlighted by all the maps generated by the directional filters. As a result, all the edge mappings of various sizes and orientations are included within the feature map, and the very best weighted pixels emerge from the final map.

Localization: dilation of image: the process of localization entails improving text regions while removing non-text areas. One of the characteristics of text is that all characters in the image usually appear near each other, forming a cluster. These prospective text pixels can be grouped using a morphological dilation technique, eliminating pixels too far from the candidate text areas. Dilation is an operation that uses a structural element of the required form and size to extend or enhance the region of interest. A vital structural element is used in the dilation process to improve the sections near one other. A structuring element with a size of 7×7 is utilized in this method. After dilatation, the resulting picture may contain non-text areas or noise that must be removed. To get rid of the noisy blobs in the image, area-based filtering is used. Only areas larger than or equal to 1/20 of the maximum area region are kept in the final picture.

Character extraction: in this step, the input image must be made suitable for character processing and identification. The text and backdrop should be monochromatic, with high contrast between the background and the text. As a result, this technique produces an image with white lettering on a black background. The algorithm's input and the final output images are displayed in figure 15.4.

15.2.2 Connected-component-based extraction

The following steps are the fundamental phases of the connected-component-based extraction algorithm, as depicted in figure 15.5.

Preprocessing: the input image is preprocessed to make the detection of text areas simpler. Only the luminance (Y) channel is required for further processing once the picture is transformed into the YUV color space (luminance and chrominance). The conversion is performed using the MATLAB function rgb2ycbcr, which takes an RGB picture as its input and transforms it into a YUV image. This updated picture may be used to extract the separate channels. The Y channel represents the image's brightness or intensity, whereas the U and V channels represent the color data. Because the text in a picture has high contrast with its backdrop, it may be converted to a gray scale image by using only the Y channel and the brightness/contrast information.

Detection of edges: To distinguish possible text areas from non-text sections, this system employs the connected-component-based method. A weight is allocated to

Figure 15.4. Input image and final image (edge based).

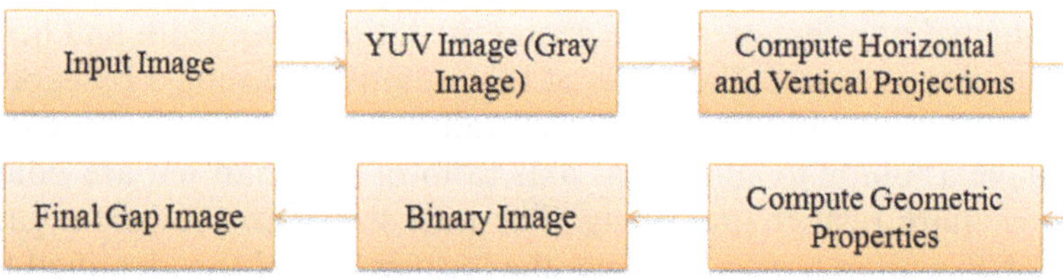

Figure 15.5. Flow chart (connected-component-based extraction).

each pixel within the edge image, which represents the count of its neighbors in specified directions. This weighting is calculated according to the maximum value of the pixel and its neighbors in the left (L), up (U), and upper-right (UR) directions. The suggested technique uses these three neighbor values to spot horizontal, vertical, and diagonal edges. The algorithm used to compute the edge image I is as follows:

1. *Set Left, Upper, UpperRight to 0.*
2. *Do GI(x,y) for all the pixels in the gray image.*
   - $Left = (GI(x,y)\ GI(x-1,y))$
   - $Upper = (GI(x,y)\ GI(x,y-1))$
   - $Upper\ Right = (GI(x,y)-GI(x+1,y-1))$
   - $I(x,y) = max\ value\ (Left,\ Upper,\ UpperRight)$
3. *Convolute the picture I using a sharpening filter to make it sharper.*
   $Weight(x,y) = max\ value(Left,\ Upper,\ UpperRight)$

The resulting edge picture is sharpened to improve the contrast between the identified edges and their backdrop, making text extraction simpler.

Localization: in this step, we examine the horizontal and vertical projection profiles for the potential text regions of the image. The sharpened edge image is used as the input intensity image to compute projection profiles with white candidate text areas on a black background. All the pixels in each intensity column of the sharpened image are shown within the vertical projection profile. In the same way, the horizontal projection profile displays the entire number of pixels in each row of the intensity image. Each bin represents the total number of pixels in each row or column of the projection profiles, which are histograms. To partition the candidate text regions, adaptive threshold values, T_y and T_x, are computed for the vertical and horizontal projections, respectively. Only areas that fall into the required thresholds are considered to be text candidates. The value of the threshold T_y is chosen to eliminate points with a robust vertical orientation, such as doors, window edges, and so on. Similarly, the value of the threshold Tx is chosen to remove areas in the horizontal direction that are not text or that have long edges:

$$T_x = \frac{\text{Mean(Horizontal projection profile)}}{20}$$

$$T_y = \text{Mean(Vertical projection profile)} + \frac{\text{Mean(Vertical projection profile)}}{10}.$$

Enhancement and gap filling: image enhancement is performed in this step. To eliminate potential non-text areas, the geometric ratio of the width and height of the text characters is estimated. The value of this ratio is determined by experimenting with various pictures to obtain an average value. In this chapter, we consider that areas that have a minor to significant axis ratio of less than ten are potential text regions that require further processing. The next step is to produce a gap picture, which is used as a reference to fine-tune the localization of the identified text areas. Assume that a pixel in the binary edge picture is bordered in all directions by black (background) pixels: vertically, horizontally, and diagonally. In this case, the background value is likewise replaced for this pixel. This procedure is known as 'gap filling.' An input picture and the final image retrieved using these techniques are shown in figure 15.6.

15.2.3 Histogram-based edge processing

As illustrated in figure 15.7, the working approach comprises six major steps.

Preprocessing: this method works with any image, regardless of format. A two-dimensional gray scale picture is required. Therefore, for a 3D input image, it is important to first convert it into a gray scale version before passing it to the next stage.

Filtering: the next step is dilation. In dilation, the edges of the white-pixel regions are gradually morphologically expanded, resulting in smaller gaps in the image. Additionally, dilation helps in filtering out different distortions introduced while converting an image to its gray scale form, viz. insufficient sharpness, color-specific fluctuations, glitter, intensity, and so on. A median filter is applied to the dilated image, which minimizes distortions and eliminates 'salt and pepper' noise. The kernel size is set to 3×3 for this step [5].

Horizontal and vertical edge processing: the column-wise and row-wise histograms are respectively represented by a horizontal histogram and a vertical histogram, both of which indicate the overall gray-value variations between consecutive pixels in a picture. The successive difference between two pixels in a column is used to

Figure 15.6. An input picture and the final image (connected-component-based extraction).

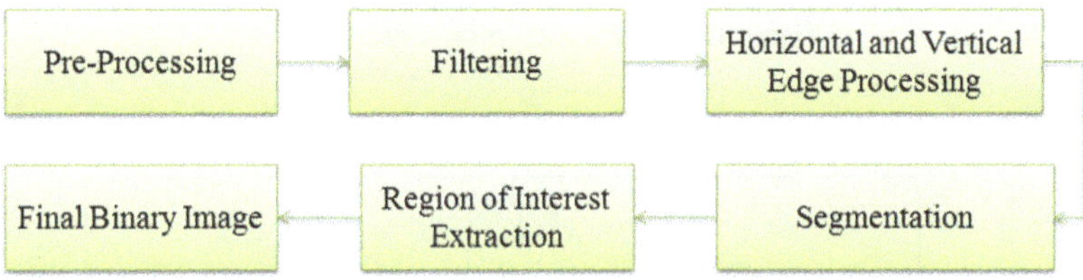

Figure 15.7. Flow chart (histogram-based edge processing).

generate the horizontal histogram. This variation is additionally appended to the existing sum total of the differences, if it reaches a specific threshold. At the end of this process, the column-wise total is stored in an array. The vertical histogram is calculated similarly, except that, instead of columns, rows are taken into account. Histograms are smoothed with a low-pass filter to prevent the loss of crucial information. After the histograms are smoothed, a final digital filter eliminates any undesirable regions from the image. If the histogram value of a row or a column is less than a particular threshold (which is computed by taking the mean of the whole histogram), then the corresponding row or column is treated as redundant. This approach produces a histogram that displays areas with a high chance of having a number plate [5]. The horizontal and vertical projection profiles are depicted in figures 15.8 and 15.9, respectively.

Segmentation: the processed picture is segmented in this phase, and the pixel values with a high probability of containing the number plate are saved in a vector.

Region of interest (ROI) extraction: in order to extract the common ROI, the previously segmented areas are processed both row- as well as column-wise by finding the largest histogram values (horizontal and vertical). The license plate is shown prominently in this region. The input and final images are presented in figure 15.10.

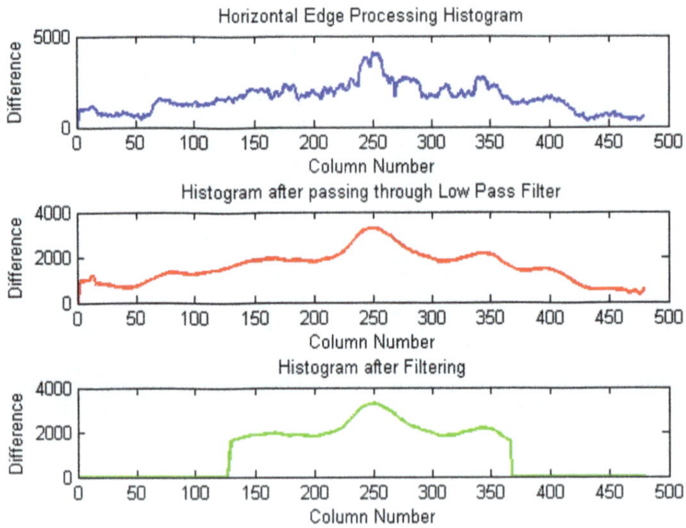

Figure 15.8. Horizontal projection profiles.

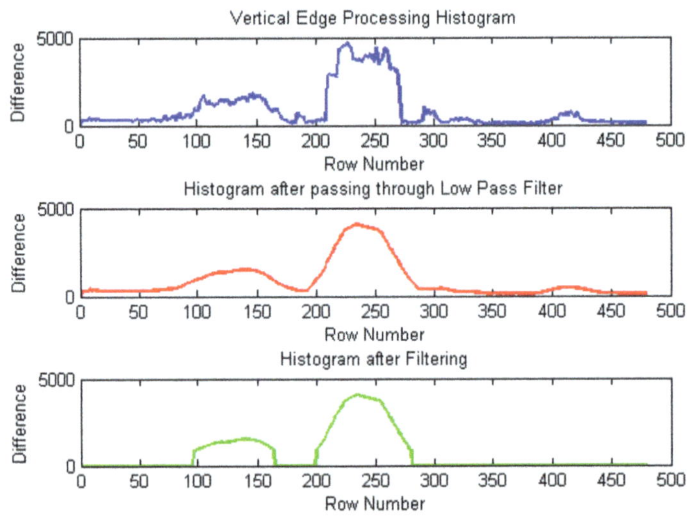

Figure 15.9. Vertical projection profiles.

15.3 Comparative results

Table 15.1 summarizes the benefits and drawbacks of each of the three methods.

To simulate all three approaches, a database of 100 unique images was employed. The average recall rate, the accuracy rate and the simulation time required by each method were used to assess the performance. The accuracy rate is obtained by dividing the number of words that were properly identified by the number of words that were successfully detected plus the number of false positives. We divide the number of successfully identified words by the number of true positive words plus

Figure 15.10. Input image and final image (histogram-based processing).

Table 15.1. Advantages and disadvantages.

Algorithm	Advantages	Disadvantages
Edge based	• More resilient in low-light environments • Adaptable for varying scales	• Does not take into account the reduction of noise prior to calculation
Connected-component based	• More resilient in low-light environments • In comparison to edge-based text extraction, it is invariant to acceptable scale changes.	• Does not take into account the reduction of noise prior to calculation
Histogram based	• Complexity is greatly reduced. • Swift compared to the methods above • Independent of the image's color type	• Number plate must be horizontally aligned and can only have a slight inclination from the horizontal axis (about 8–10 degrees).

the number of false negatives to get the recall rate. False negatives are image domains with text characters that the algorithm does not recognize [5]. The rates of precision and recall are mathematically defined as follows:

$$\text{Precision rate} = \frac{\text{Successfully detected}}{\text{Successfully detected} + \text{False positives}}$$

$$\text{Recall rate} = \frac{\text{Successfully detected}}{\text{Successfully detected} + \text{False negatives}}$$

Table 15.2 shows the accuracy and recall rates and the MATLAB simulation time required when each technique was implemented and evaluated for 100 pictures. Table 15.3 also shows the average run duration and success rate for various environmental conditions. Based on these results, it can be concluded that the histogram-based edge-processing method is efficient and easy, and that it can definitely be employed for an FPGA implementation of a VLPDS.

15.4 Hardware–software co-simulation of automatic license plate detection on the ZedBoard SoC platform

The hardware implementation is made up of two components: (1) the MATLAB Simulink model and (2) the XSG model. Some blocks must be constructed manually using 'Vivado HLS,' a high-level synthesis environment. To calculate the histogram values, some code has to be written in C. After being converted to register-transfer level (RTL), this code may be used as a block in System Generator.

15.4.1 Simulink model

First, we created a MATLAB Simulink model, as illustrated in figure 15.11, for hardware implementation.

15.4.1.1 Simulink blocks used for the implementation

Image from file: 'Image from File' is the initial block, which imports images from the database. This block specifies the name of the graphics file containing the picture to be imported into the Simulink modeling and simulation program using the

Table 15.2. Precision and recall rates.

Algorithm	Precision rate (%)	Recall rate (%)	MATLAB simulation time (s)
Edge based	91	96.7	~20
Connected-component based	94	95.2	~17
Histogram based	95	97.8	~11

Table 15.3. Success rate under different environmental conditions.

Algorithm	License plate conditions	Success rate (%)	Average run time (s)
Edge-based extraction	• Sunlight	73	~19
Connected-component-based extraction	• Cloudy weather • Shade	80	~23
Histogram-based edge processing	• Different colored backgrounds	96	~14

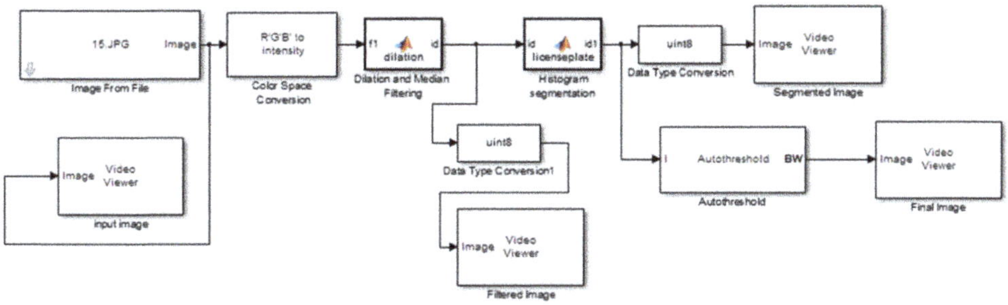

Figure 15.11. Simulink model of the VLPDS.

'Filename' parameter. The Browse button can be used to locate the file if it is not on the MATLAB path. This parameter supports URL paths.

Color-space conversion: this block converts an RGB color space image into the intensity and YCbCr color spaces, and vice versa. This can also be used to convert a color image to a gray scale image for simplicity and high processing speed.

Dilation and median filtering: we can write a MATLAB function for use in a Simulink model using a MATLAB function block. We wrote a MATLAB function that simulates and creates code for a Simulink Coder target. We utilize this block as a 'dilation' block, creating a user-defined function using MATLAB programming. The resultant picture is then dilated before being smoothed with a median filter.

License plate: the next block, 'license plate,' is another user-defined function that combines the algorithm's histogram calculation and segmentation phases.

Auto threshold: the 'Autothreshold' block is used to transform an intensity image into a binary image using an Otsu's method threshold value.

Data type conversion: The 'Data type conversion' block converts the final picture into unsigned integer format. It transforms the input to the output's data type and size. In this conversion, the input and output have the same real-world values, and the input and output have the same stored value.

Video viewer: The 'Video Viewer' section displays the output of any stage. It shows pictures or video streams in the binary, intensity, or RGB formats. When the model is running, the video viewer block provides simulation controls to allow the user to play, pause, and step. This block also contains tools for analyzing pixel regions. The Simulink block does not produce code for this block during code generation.

15.4.2 System generator model

MATLAB Simulink and Xilinx Vivado were used to implement the hardware model in XSG. The Xilinx system generator token must be set up using MATLAB, which adds the Xilinx block set to the MATLAB Simulink environment. This block set may be used to create an algorithmic model right away. The XSG is connected to a suitable FPGA board once the required results have been obtained. The model is implemented through JTAG hardware co-simulation on a high-capacity ZedBoard, the Zynq evaluation and development FPGA board based on the ZYnq series

XC7Z020–1CLG484. Figure 15.12 depicts an FPGA-based VLPDS architecture developed in Simulink using the Xilinx block set.

15.4.2.1 System generator blocks used for the implementation

Preprocessing: this model's starting blocks are the same as those of the Simulink model mentioned above.

Unbuffer: following the filtering, the 'Unbuffer' block helps to deliver the picture pixels serially. It outputs scalar samples at a faster pace after converting a frame. This block converts an M-by-N input to a 1-by-N output by unbuffering it. In other words, inputs are unbuffered row by row, such that each matrix row in the output becomes a particular time sample.

Gateway blocks: The FPGA boundaries were initially defined using the gateway blocks 'Gateway In' and 'Gateway Out,' in which the algorithmic model has to be built using Xilinx system blocks. The 'Gateway In' block transforms Simulink integer, single, double, and fixed-point data types to Xilinx fixed-point and floating-point data types. This block is used to supply fixed-point data and receives image data streams. These blocks become top-level input ports in the hardware. The 'Gateway Out' block, on the other hand, transforms Xilinx fixed-point or floating-point inputs to Simulink integer, single, double, or fixed-point outputs. These blocks become top-level output ports in the hardware. Figure 15.13 shows a snapshot of the gateway blocks present in the XSG.

System generator token: The floating system generator token is inserted into the model and specifies how the model should be burned into the hardware. Here, we used the System Generator token to provide accelerated simulation through hardware co-simulation.

Vivado high-level synthesis (HLS): Xilinx's Vivado HLS software reconstructs a C specification within an RTL implementation, which is subsequently synthesized into an FPGA and an SoC. C specifications can be written in C, C++, or System-C. HLS unites the software and hardware systems. HLS:

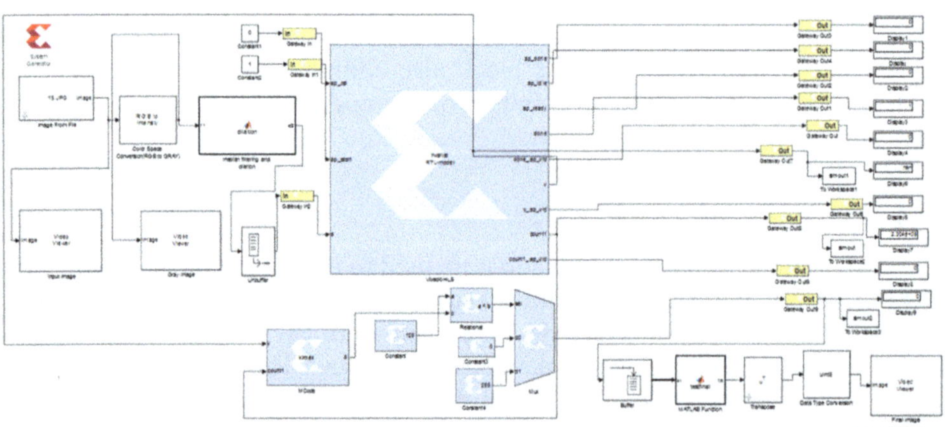

Figure 15.12. Hardware–software co-simulation model for a VLPDS.

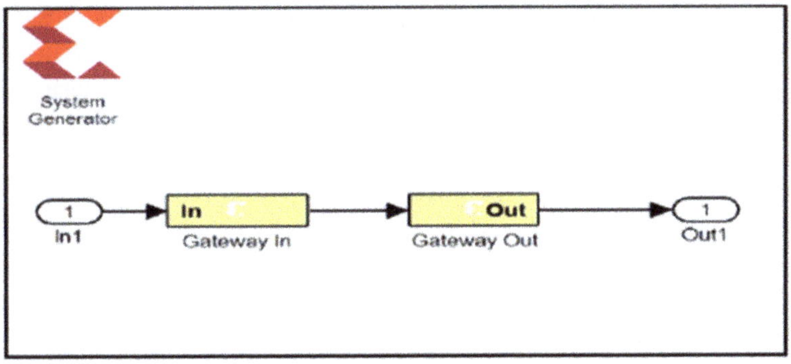

Figure 15.13. Gateway blocks.

- allows FPGA-based hardware designers to get the benefit of the enhanced effectiveness that comes from working at a higher level of abstraction when designing high-performance hardware
- empowers programmers to easily simulate computationally complex blocks of their proposed algorithms on a new compilation target. The FPGA is a massively parallel architecture that transcends standard processors in performance, cost, and power.

As previously stated, the HLS environment 'Vivado HLS' was utilized to develop the blocks that were unavailable in XSG. Both the histogram calculation as well as the segmentation procedure are coded in C. It is now synthesized in the HLS environment, yielding RTL that can be used to construct blocks in the XSG environment.

Algorithm for thresholding: In the final stage, the intensity picture must be transformed back into a binary image. When an image is partitioned into foreground and background, all pixels above a certain threshold become white, while the rest become black. The 'Relational' block is utilized to compare two values, while the 'MUX' and 'Constant' (128 in this case) blocks are used to replace the thresholds with white.

Buffer: Finally, the 'buffer' block is used in postprocessing to redistribute the data in each input column in order to generate the output. This slows down the conversion of scalar samples to a frame output. With an adjustable overlap, we may also transform a frame to a smaller or bigger size. This block performs frame-based processing at all times. The block redistributes the data in each input column to generate output with a different frame size. When a signal is buffered to create a larger frame, the output is slower than the input. The user can view the vehicle's license plate in the 'Video Viewer' block.

JTAG hardware co-simulation: The XSG is a powerful tool for executing image processing operations in both software and hardware. XSG provides quick hardware implementation of advanced image processing algorithms with limited resources and low latency as it restricts the usage of design processing units. One can see the netlist generated during compilation, as well as a model and a programming file in VHDL, which are accessible via Xilinx ISE and Xilinx Vivado. Finally, the JTAG hardware co-simulation is created for the ZedBoard (a Zynq evaluation and development FPGA board with the ZYnq series XC7Z020–1CLG484). The Xilinx Zynq-7000 extensible processing platform (EPP) is the basis for the ZedBoard, which is an evaluation and development board. The Zynq-7000 EPP combines a twin Corex-A9 processing system (PS) with 85,000 Series-7 programmable logic (PL) cells, making it suitable for a wide range of applications. The ZedBoard's diverse set of onboard peripherals and extension options make it an excellent platform for both new and seasoned designers. Figure 15.14 depicts the JTAG hardware co-simulation block.

15.5 Results and utilization summary

We examine the resulting compilation netlist and build a VHDL programming file that can be accessed and run via Xilinx ISE and Vivado. The module may then be tested for behavioral syntax error correction, synthesized, and programmed into an FPGA. The XSG can generate a user constraints file (UCF), a test bench, and test vectors for architectural testing. Bitstream compilation is then completed, which is required to produce an FPGA bit file that can be used for FPGA input. We utilized the 'hardware co-simulation' option forthis study (JTAG). The System Generator model or subsystem we want to execute in hardware is the starting point for hardware co-simulation after our hardware board is mounted. A model can be co-simulated if it fulfills the hardware board's criteria. Figures 15.15 and 15.16 show snapshots of the hardware–software co-simulation implementation of VLPDS on the ZedBoard.

15.5.1 Resource utilization summary

The resource usage and latency data for the Vivado HLS block, which implements the algorithm, are shown in table 15.4. This table shows the random access memory (RAM) used (18K of block RAM (BRAM)), the digital signal processing (DSP48E), the flip-flops (FF), and the look-up table (LUT). This table summarizes the use of the ZedBoard, a Zynq evaluation and development FPGA board from the XC7Z020–1CLG484 Zynq family. It makes use of half of the on-chip resources but just 7% of the LUTs, resulting in an accuracy of 99.3%. At a frequency of 100 MHz, it can process one image (480 × 480 pixels) in 2.7 ms.

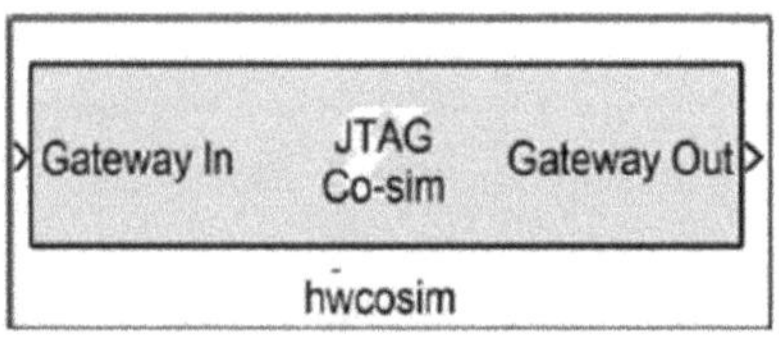

Figure 15.14. JTAG block.

Figure 15.15. The bitstream file is configured on the ZedBoard.

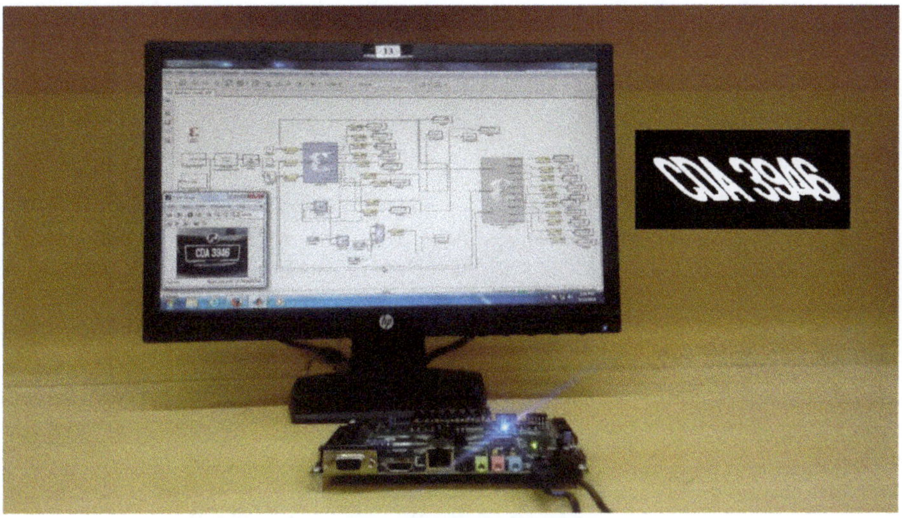

Figure 15.16. Hardware–software co-simulation of VLPDS.

Table 15.4. Resource utilization summary.

Name	BRAM_18K	DSP48E	FF	LUT
Expression		3	0	2782
FIFO				
Instance		14	0	0
Memory	98		40	4
Multiplexer				1136
Register			1930	
Total	98	17	1970	3922
Available	280	220	106400	53200
Utilization (%)	35	7	1	7

15.6 Conclusions and future work

Because of time-to-market constraints, quick prototyping tools, such as MATLAB Simulink and XSG have grown increasingly essential in recent years, according to the literature. They simplify the hardware implementation process. These programs can mimic software, but their most useful feature is their ability to generate FPGA hardware; parallelism, resilience, and speed are all important characteristics. In this study, we built an FPGA-based hardware version of a VLPDS. Vivado HLS was used to manually generate certain blocks using code written in the C computer language. These blocks can be accessed in XSG and further processed. The XSG model's accuracy is higher than 80%, allowing for fast hardware co-simulation and the implementation of advanced image processing algorithms with few resources and little delay. On the other hand, if we increase the size of the image, the number of resources used will also increase. However, we can easily implement the proposed system architecture on any FPGA board using a digital signal processor that is compatible with Vivado HLS. In the future, we want to include a digital camera and a webcam as part of the design to make the system more practical. In the future, we can also add a security mechanism to VLPDS for better efficacy and practicability.

Bibliography

[1] Shafi I *et al* 2021 License plate identification and recognition in a non-standard environment using neural pattern matching *Complex Intell. Syst.* pp 1–13

[2] Anagnostopoulos C-N E *et al* 2008 License plate recognition from still images and video sequences: A survey *IEEE Trans. Intell. Transp. Syst.* **9.3** 377–91

[3] Mufti N and Shah S A A 2021 Automatic number plate Recognition: A detailed survey of relevant algorithms *Sensors* **21.9** 3028

[4] Akpojotor P *et al* 2021 Automatic license plate recognition on microprocessors and custom computing platforms: A review *IET Image Processing*

[5] Chhabra S, Jain H and Saini S 2016 FPGA based hardware implementation of automatic vehicle license plate detection system *2016 Int. Conf. on Advances in Computing, Communications and Informatics (ICACCI)* (Piscataway, NJ: IEEE)

[6] Jung K, Kim K I and Jain A K 2004 Text information extraction in images and video: a survey *Pattern Recognit.* **37.5** 977–97

[7] Al-Ghaili A M *et al* 2008 A new vertical edge detection algorithm and its application *2008 International Conference on Computer Engineering & Systems* (IEEE) pp 204–9

[8] Lienhart R and Wernicke A 2002 Localizing and segmenting text in images and videos *IEEE Trans. Circuits Syst. Video Technol.* **12.4** 256–68

[9] Wu V, Manmatha R and Riseman E M 1999 Textfinder: An automatic system to detect and recognize text in images *IEEE Trans. Pattern Anal. Mach. Intell.* **21.11** 1224–9

[10] Ye Q *et al* 2003 A robust text detection algorithm in images and video frames *Fourth Int. Conf. on Information, Communications and Signal Processing, 2003 and the Fourth Pacific Rim Conf. on Multimedia. Proc. of the 2003 Joint* 2 (Piscataway, NJ: IEEE) pp 802–6

[11] Liu X and Samarabandu J K 2005 A simple and fast text localization algorithm for indoor mobile robot navigation *Proc. SPIE* 5672 *Image Processing: Algorithms and Systems IV*

[12] Wang K and Kangas J A 2003 Character location in scene images from digital camera *Pattern Recognit.* **36.10** 2287–99

[13] Kim K C *et al* 2004 Scene text extraction in natural scene images using hierarchical feature combining and verification *Proc. 17th Int. Conf. on Pattern Recognition, 2004. ICPR 2004* 2 (Piscataway, NJ: IEEE) pp 679–82

[14] Giri a S 2013 Text information extraction and analysis from images using digital image processing techniques *Special Issue of International Journal on Advanced Computer Theory and Engineering (IJACTE)* **2.1** 66–71

[15] Das M, Swamy B, Bindhu H and Govardhan A 2012 Evaluation of text detection and localization methods in natural images *Int. J. Emerging Technology and Advanced Engineering* **2.6** 277–82

[16] Angadi S A and Kodabagi M M IEEE Text region extraction from low resolution natural scene images using texture features *2010 IEEE 2nd Int. Advance Computing Conf. (IACC)* (IEEE) pp 121–8

[17] Lee C-M and Kankanhalli A 1995 Automatic extraction of characters in complex scene images *Int. J. Pattern Recognit Artif Intell.* **9.01** 67–82

[18] Gupta P, Purohit G N and Rathore M 2014 Number plate extraction using template matching technique *International Journal of Computer Applications* **88.3** 40–4

[19] Chunyu C *et al* 2013 Application of image processing to the vehicle license plate recognition *2nd International Conference on Computer Science and Electronics Engineering (ICCSEE 2013)* (Atlantis Press) pp 2867–9

[20] Khalifa O O *et al* 2007 Malaysian Vehicle License Plate Recognition *Int. Arab J. Inf. Technol.* **4** 359–64

[21] Sulaiman N *et al* 2013 Development of automatic vehicle plate detection system *IEEE 3rd Int. Conf. on System Engineering and Technology*

[22] Liu X and Samarabandu J 2005 An edge-based text region extraction algorithm for indoor mobile robot navigation *IEEE Int. Conf. Mechatronics and Automation, 2005* vol 2 (Piscataway, NJ)

[23] Liu X and Samarabandu J 2006 Multiscale edge-based text extraction from complex images *2006 IEEE Int. Conf. on Multimedia and Expo*

[24] Gllavata J, Ewerth R and Freisleben B 2003 A robust algorithm for text detection in images *3rd Int. Symp. on Image and Signal Processing and Analysis, 2003. ISPA 2003* vol 2 pp 611–16

[25] Saidani T *et al* 2009 Hardware co-simulation for video processing using xilinx system generator *Proc. of the World Congress on Engineering* vol 1

[26] Karthigaikumar P, Jaraline Kirubavathy K and Baskaran K 2011 FPGA based audio watermarking—Covert communication *Microelectron. J.* **42.5** 778–84

[27] Moreo A T *et al* 2005 Experiences on developing computer vision hardware algorithms using Xilinx system generator *Microprocess. Microsyst.* **29.8-9** 411–19

[28] Chauhan B, Jain A, Chaturvedi T and Saini S 2015 User interactive and assistive fleet management and eco-driving system *2015 IEEE Region 10 Symp.* pp 41–4

[29] Gnanaprakash V, Kanthimathi N and Saranya N 2021 Automatic number plate recognition using deep learning *IOP Conf. Ser.: Materials Sci. Eng.* **1084** 012027

www.ingramcontent.com/pod-product-compliance
Ingram Content Group UK Ltd.
Pitfield, Milton Keynes, MK11 3LW, UK
UKHW050249150926
8556IPUK00050B/697